# 3D PRINTING IN SPACE

Committee on Space-Based Additive Manufacturing

Aeronautics and Space Engineering Board

National Materials and Manufacturing Board

Division on Engineering and Physical Sciences

NATIONAL RESEARCH COUNCIL
*OF THE NATIONAL ACADEMIES*

THE NATIONAL ACADEMIES PRESS
Washington, D.C.
**www.nap.edu**

**THE NATIONAL ACADEMIES PRESS** **500 Fifth Street, NW** **Washington, DC 20001**

NOTICE: The project that is the subject of this report was approved by the Governing Board of the National Research Council, whose members are drawn from the councils of the National Academy of Sciences, the National Academy of Engineering, and the Institute of Medicine. The members of the committee responsible for the report were chosen for their special competences and with regard for appropriate balance.

This report is based on work supported by Contract NNH10CD04B (Task Order 7) between the National Academy of Sciences and the National Aeronautics and Space Administration and Grant FA9453-11-3-0001 between the National Academy of Sciences and the United States Air Force. Any opinions, findings, conclusions, or recommendations expressed in this publication are those of the authors and do not necessarily reflect the views of the agencies that provided support for the project.

International Standard Book Number-13: 978-0-309-31008-6
International Standard Book Number-10: 0-309-31008-3

*Cover:* Design by Tim Warchocki.

Copies of this report are available free of charge from

Aeronautics and Space Engineering Board
National Research Council
The Keck Center of the National Academies
500 Fifth Street, NW
Washington, DC 20001

Additional copies of this report are available from the National Academies Press, 500 Fifth Street, NW, Keck 360, Washington, DC 20001; (800) 624-6242 or (202) 334-3313; http://www.nap.edu.



Printed in the United States of America

# THE NATIONAL ACADEMIES

*Advisers to the Nation on Science, Engineering, and Medicine*

The **National Academy of Sciences** is a private, nonprofit, self-perpetuating society of distinguished scholars engaged in scientific and engineering research, dedicated to the furtherance of science and technology and to their use for the general welfare. Upon the authority of the charter granted to it by the Congress in 1863, the Academy has a mandate that requires it to advise the federal government on scientific and technical matters. Dr. Ralph J. Cicerone is president of the National Academy of Sciences.

The **National Academy of Engineering** was established in 1964, under the charter of the National Academy of Sciences, as a parallel organization of outstanding engineers. It is autonomous in its administration and in the selection of its members, sharing with the National Academy of Sciences the responsibility for advising the federal government. The National Academy of Engineering also sponsors engineering programs aimed at meeting national needs, encourages education and research, and recognizes the superior achievements of engineers. Dr. C. D. Mote, Jr., is president of the National Academy of Engineering.

The **Institute of Medicine** was established in 1970 by the National Academy of Sciences to secure the services of eminent members of appropriate professions in the examination of policy matters pertaining to the health of the public. The Institute acts under the responsibility given to the National Academy of Sciences by its congressional charter to be an adviser to the federal government and, upon its own initiative, to identify issues of medical care, research, and education. Dr. Victor J. Dzau is president of the Institute of Medicine.

The **National Research Council** was organized by the National Academy of Sciences in 1916 to associate the broad community of science and technology with the Academy's purposes of furthering knowledge and advising the federal government. Functioning in accordance with general policies determined by the Academy, the Council has become the principal operating agency of both the National Academy of Sciences and the National Academy of Engineering in providing services to the government, the public, and the scientific and engineering communities. The Council is administered jointly by both Academies and the Institute of Medicine. Dr. Ralph J. Cicerone and Dr. C. D. Mote, Jr., are chair and vice chair, respectively, of the National Research Council.

**www.nationalacademies.org**

**OTHER REPORTS OF THE AERONAUTICS AND SPACE ENGINEERING BOARD**

Autonomy Research for Civil Aviation: Toward a New Era of Flight (Aeronautics and Space Engineering Board [ASEB], 2014)
Pathways to Exploration: Rationales and Approaches for a U.S. Program of Human Space Exploration (ASEB with Space Studies Board [SSB], 2014)

Solar and Space Physics: A Science for a Technological Society (SSB with ASEB, 2013)

Continuing Kepler's Quest: Assessing Air Force Space Command's Astrodynamics Standards (ASEB, 2012)
NASA Space Technology Roadmaps and Priorities: Restoring NASA's Technological Edge and Paving the Way for a New Era in Space (ASEB, 2012)
NASA's Strategic Direction and the Need for a National Consensus (Division on Engineering and Physical Sciences, 2012)
Recapturing NASA's Aeronautics Flight Research Capabilities (ASEB, 2012)
Reusable Booster System: Review and Assessment (ASEB, 2012)

Limiting Future Collision Risk to Spacecraft: An Assessment of NASA's Meteroid and Orbital Debris Programs (ASEB, 2011)
Preparing for the High Frontier—The Role and Training of NASA Astronauts in the Post-Space Shuttle Era (ASEB, 2011)
Recapturing a Future for Space Exploration: Life and Physical Sciences Research for a New Era (SSB and ASEB, 2011)

Advancing Aeronautical Safety: A Review of NASA's Aviation Safety-Related Research Programs (ASEB, 2010)
Capabilities for the Future: An Assessment of NASA Laboratories for Basic Research (Laboratory Assessments Board with SSB and ASEB, 2010)
Defending Planet Earth: Near-Earth Object Surveys and Hazard Mitigation Strategies (SSB with ASEB, 2010)
Forging the Future of Space Science: The Next 50 Years: An International Public Seminar Series Organized by the Space Studies Board: Selected Lectures (SSB with ASEB, 2010)

America's Future in Space: Aligning the Civil Space Program with National Needs (SSB with ASEB, 2009)
Approaches to Future Space Cooperation and Competition in a Globalizing World: Summary of a Workshop (SSB with ASEB, 2009)
An Assessment of NASA's National Aviation Operations Monitoring Service (ASEB, 2009)
Fostering Visions for the Future: A Review of the NASA Institute for Advanced Concepts (ASEB, 2009)
Near-Earth Object Surveys and Hazard Mitigation Strategies: Interim Report (SSB with ASEB, 2009)
Radioisotope Power Systems: An Imperative for Maintaining U.S. Leadership in Space Exploration (SSB with ASEB, 2009)

Limited copies of ASEB reports are available free of charge from

Aeronautics and Space Engineering Board
National Research Council
The Keck Center of the National Academies
500 Fifth Street, NW, Washington, DC 20001
(202) 334-2858/aseb@nas.edu
www.nationalacademies.org/aseb

## COMMITTEE ON SPACE-BASED ADDITIVE MANUFACTURING

ROBERT H. LATIFF, R. Latiff Associates, *Chair*
ELIZABETH R. CANTWELL, Lawrence Livermore National Laboratory, *Vice Chair*
PETER M. BANKS, Red Planet Capital Partners
ANDREW S. BICOS, The Boeing Company
RAVI B. DEO, EMBR
JOHN W. HINES, Senior Technology Advisor, Independent Consultant
BHAVYA LAL, IDA Science and Technology Policy Institute
SANDRA H. MAGNUS, American Institute of Aeronautics and Astronautics
THOMAS E. MAULTSBY, Rubicon, LLC
MICHAEL T. McGRATH, University of Colorado, Boulder
LYLE H. SCHWARTZ, Air Force Office of Scientific Research (Retired)
IVAN E. SUTHERLAND, Portland State University
RYAN WICKER, University of Texas, El Paso
PAUL K. WRIGHT, Berkeley Energy and Climate Institute, University of California, Berkeley

***Staff***

DWAYNE A. DAY, Senior Program Officer, Aeronautics and Space Engineering Board
ERIK B. SVEDBERG, Senior Program Officer, National Materials and Manufacturing Board
ANDREA REBHOLZ, Program Associate, Aeronautics and Space Engineering Board

MICHAEL H. MOLONEY, Director, Aeronautics and Space Engineering Board and Space Studies Board

## AERONAUTICS AND SPACE ENGINEERING BOARD

## NATIONAL MATERIALS AND MANUFACTURING BOARD

# Preface

Additive manufacturing, often referred to as "3D printing," has attracted significant attention recently, including discussion of its applications to spaceflight. NASA, the Air Force Space Command, and the Air Force Research Laboratory asked the National Research Council (NRC) to conduct a study of the prospects for the use of additive manufacturing in space. In response, the NRC established the Committee on Space-Based Additive Manufacturing. The committee's statement of task required it to

- Assess the current state of additive manufacturing in the United States and worldwide (especially in the aerospace industries, universities, and national laboratories engaged in the design and manufacture of small satellites or respective subassemblies);
- Characterize the future states envisioned by the aerospace industries, universities, and national laboratories with respect to additive manufacturing and aerospace systems;
- Discuss the feasibility of the concept of space-based additive manufacturing of space hardware (including, but not limited to, a fully functional small spacecraft) that can conduct or enable missions of relevance to NASA, the Air Force, and/or the national security space communities;
- Identify the science and technology gaps between current additive manufacturing capabilities and the capabilities required to enable a space-based additive manufacturing concept, including those gaps that current trends indicate may be closed with commercial investments in additive manufacturing and those gaps that are likely to require dedicated investments by the federal government.
- Assess the implications that a space-based additive manufacturing capability would have on launch requirements (e.g., launching raw materials versus fully assembled spacecraft); overall satellite and payload designs; and in-space operations, such as possible reductions in mass and their implications for activities such as maneuverability.

The first two tasks are respectively addressed in Chapters 1 and 2 of this report. The remaining three are addressed in Chapter 3. Rather than arrange the chapters according to the statement of task, the committee devotes Chapter 4 to NASA issues and Chapter 5 to Air Force issues, while noting that both the Air Force and NASA can benefit from coordinating their efforts in developing this technology. Particularly in Chapter 3 the committee identified many of the challenges that have to be overcome and the issues that have to be taken into consideration in order to use the technology in space. The committee noted that although commercial investment in ground-based additive manufacturing for aerospace use is extensive, the conservatism of the aerospace industry and the high costs and unclear value of in-space additive manufacturing means that the government will have to take the lead in developing this technology. In addition, because the application of this technology to in-space use is so new (as of the writing of this report, the first in-space additive manufacturing experiments were planned by the end of 2014), it is difficult to draw firm conclusions about how the technology may impact issues such as launch requirements. As the committee notes in several places (for example, Chapter 2), a benefit of this technology may not be to reduce launch mass but to enable new capabilities (i.e., satellite and payload designs).

The statement of task also stated that the committee may also consider the following:

- The potential mission payloads and capabilities that could be expected from a space-based, additively manufactured spacecraft;
- The role in potential missions for a single spacecraft system manufactured in space by additive manufacturing or for multiple spacecraft systems, including disaggregated constellations and fractionated satellites;
- Concepts of operations for space-based manufacture of space hardware (including small spacecraft) using additive manufacturing, including development, test and evaluation, launch, deployment, and on-orbit command and control;
- Whether it is possible to develop a high-level heuristic tool that Air Force Space Command and other government organizations could use for first-order assessments of space-based, additively manufactured small spacecraft concepts in their integrated planning and process efforts.

Possible future applications of the technology are particularly addressed in Chapter 2. The committee notes that the value of this technology will be demonstrated in the nearer term at the component level rather than the manufacture of entire spacecraft. In Chapters 4 and 5, it recommends that as the technology develops, NASA and the Air Force both apply cost-benefit analysis to the technology but also recognize that new capabilities (i.e., the benefits) should not be ignored.

# Acknowledgment of Reviewers

This report has been reviewed in draft form by individuals chosen for their diverse perspectives and technical expertise, in accordance with procedures approved by the National Research Council's (NRC's) Report Review Committee. The purpose of this independent review is to provide candid and critical comments that will assist the institution in making its published report as sound as possible and to ensure that the report meets institutional standards for objectivity, evidence, and responsiveness to the study charge. The review comments and draft manuscript remain confidential to protect the integrity of the deliberative process. We wish to thank the following individuals for their review of this report:

James B. Armor, ATK, Spacecraft System & Services,
Joseph J. Beaman, University of Texas, Austin,
Mary Anne Fox, University of California, San Diego,
Sven Grahn, Swedish Space Corporation (retired),
Douglas C. Hofmann, NASA Jet Propulsion Laboratory/California Institute of Technology,
Kevin Jurrens, National Institute of Standards and Technology,
Eric MacDonald, University of Texas, El Paso,
Ted Nye, California State University, and
Christopher M. Spadaccini, Lawrence Livermore National Laboratory.

Although the reviewers listed above have provided many constructive comments and suggestions, they were not asked to endorse the conclusions or recommendations, nor did they see the final draft of the report before its release. The review of this report was overseen by Mark C. Hersam, Northwestern University. Appointed by the NRC, he was responsible for making certain that an independent examination of this report was carried out in accordance with institutional procedures and that all review comments were carefully considered. Responsibility for the final content of this report rests entirely with the authoring committee and the institution.

# Contents

APPENDIXES

# Summary

Additive manufacturing has the potential to positively affect human spaceflight operations by enabling the in-orbit manufacture of replacement parts and tools, which could reduce existing logistics requirements for the International Space Station (ISS) and future long-duration human space missions. The benefits of in-space additive manufacturing for robotic spacecraft are far less clear, although this rapidly advancing technology can also potentially enable space-based construction of large structures and, perhaps someday, substantially in the future, entire spacecraft. Additive manufacturing can also help to reimagine a new space architecture that is not constrained by the design and manufacturing confines of gravity, current manufacturing processes, and launch-related structural stresses.

The specific benefits and potential scope of additive manufacturing remain undetermined, and there has been a substantial degree of exaggeration, even hype, about its capabilities in the short term. The public often believes that these technologies are further along than they actually are. The realities of what can be accomplished today, using this technology on the ground, demonstrate the substantial gaps between the vision for additive manufacturing in space and the limitations of the technology and the progress that has to be made to develop it for space use. What can be accomplished in the far future depends on many factors, including decisions made today by NASA and the Air Force.[1]

When looking at the potential values of in-space additive manufacturing, the Committee on Space-Based Additive Manufacturing found that ground-based additive manufacturing for aerospace systems has more immediate and long-term impacts to reduce cost and increase performance of space systems, as well as establish the technical basis of later, space-based additive manufacturing. The committee also determined that additive manufacturing in and of itself is not a solution, but presents potential opportunities, both as a tool in a broad toolkit of options for space-based activities and as a potential paradigm-changing approach to designing hardware for in-space activities.

## THE ORIGINS OF THIS STUDY

The concept of space-based manufacturing in general has existed almost since the beginning of the space age but has made limited progress because of the difficulties of space-based construction. However, additive manu-

[1] Although there are other government space actors, including the National Oceanic and Atmospheric Administration, the Navy, the National Reconnaissance Office, and so on, this study was commissioned by the Air Force and NASA and therefore focuses on their missions.

facturing, often referred to as "3D printing," has captured the public imagination in recent years and received considerable media attention. The technology is interesting and exciting, it can provide benefits over conventional manufacturing processes, and has become publicly accessible in recent years, with small desktop devices entering the market for home use. Many of the claims made in the popular press about this technology have been exaggerated, and it appears that even as it continues to advance and evolve, additive manufacturing will primarily supplement rather than replace many existing manufacturing methods.

Two sectors in particular, biomedical and aerospace, are the largest users of the technology. Additive manufacturing has potential for aerospace use to reduce costs, shorten production schedules, and enable the development of new structures. Many companies large and small are evaluating the ability of ground-based additive manufacturing to produce components for aircraft and spacecraft, and additively manufactured parts have already flown in space.

The Air Force Space Command, the Air Force Research Laboratory, and NASA's Space Technology Mission Directorate charged the National Research Council with evaluating the prospects of in-space additive manufacturing. After examining the various technologies available and in development and hearing from a wide range of experts on the subject, the committee concluded that in-space additive manufacturing is likely to have a significant impact on crewed space operations. Its potential for robotic spacecraft operations is less clear, especially in the short term. Because some of the most obvious applications are for human spaceflight, the government cannot expect private industry to sponsor space-based additive manufacturing on its own. Ground-based additive manufacturing is being rapidly developed by industry, and the committee therefore sought to determine what aspects of space-based additive manufacturing industry would not undertake on its own. The two most obvious are space-based robotics and automation and hybrid manufacturing in which two or more manufacturing processes work together, preferably in an automated way, in the space environment. Because the most obvious applications are for human spaceflight and exploration and for military missions, the government cannot expect industry to invest in technology developments that do not have a clear path to profit. The committee also determined that the ISS provides an excellent opportunity for both civilian and military research on additive manufacturing technology.

As recently as the 1990s, NASA and the Air Force as well as other military organizations, such as the Defense Advanced Research Projects Agency, conducted cooperative research with each other with substantial results. The committee believes that in-space additive manufacturing is an area where such civil-military cooperation can and should occur.

## THE PROMISE AND POTENTIAL OF SPACE-BASED ADDITIVE MANUFACTURING

Additive manufacturing as a commercial technology that builds three-dimensional parts directly from computer files has existed since the 1980s and has been evaluated for space-based use since the late 1990s. In its most basic form, additive manufacturing involves the process of adding material on top of some kind of build platform and building on it consecutively until an object is produced. This is opposed to more conventional subtractive manufacturing methods. Currently, most additive manufacturing techniques involve the use of only a single material and thus require that functional parts consisting of more than one material be developed by separate machines and undergo finishing and assembly.

The application of additive manufacturing to the space environment could likely lead to a change in our ideas and concepts of what satellites look like, how they are designed, and what they can do. Additive manufacturing is not just a different way to manufacture components and space-based devices, but rather offers a new way to reconceptualize space architectures. It enables development of structures entirely unlike those needed in the high-gravity environment of Earth or to survive the rigors of space launch. Large structures may be useful in space for many applications, from antennas to structural supports (although it is worth noting that most additive manufacturing machines today make parts smaller than themselves, so this is a different approach to the technology). Additive manufacturing can potentially lead to the construction of smaller, more reliable, less massive satellite systems or their key components (including support structure, power distribution system, solar arrays, instruments, outer protective shell, etc.), which could reduce launch requirements and costs.

The lack of gravity and atmosphere presents possibilities for additive manufacturing in space not available to ground-based machines. The absence of gravity might permit a printer to work on the "bottom" and the "top" of an object at the same time. Imagine a printer for use in space that has multiple print heads and works on all six sides of an object resting in the space between the heads. Air jets or electrostatic attraction might be used to keep the growing object in place, or even to move it to the orientation most suitable for printing. For additive manufacturing in space, considering a 20-year time horizon, NASA has a unique opportunity to encourage innovative thinking about how to capitalize on the lack of gravity or the lack of atmosphere in space to better and more rapidly form objects that are similar to those made on Earth.

Although additive manufacturing is advancing rapidly and is increasingly used on the ground for an expanding number of industrial purposes, the basic technology is still relatively young. There are some fundamental issues that industry will have to resolve before space-based applications can be derived. A clear understanding of the relationships between the material and structural properties and their dependence on processing techniques needs to be established to ensure consistency in production. The production process could also benefit from standardization of design software, file formats, and processing and equipment parameters, including developing closed-loop feedback control systems for the machines themselves. Most importantly, a verification and certification methodology will have to be defined that guarantees the quality of the additively manufactured parts.

Aerospace systems have critical missions and must meet rigorous standards for quality and reliability—standards that are set to ensure mission success. In order to benefit from additive manufacturing approaches, the manufacturing community—with government involvement—will have to address the issues of qualification and certification. A standard approach to qualification and certification of finished parts will simplify the application of additive manufacturing to the space environment and also enable more widespread application on Earth. This led the committee to its first recommendation.

**Recommendation: NASA and the Air Force should jointly cooperate—and possibly involve additional parties, including other government agencies as well as industry—to research, identify, develop, and gain consensus on standard qualification and certification methodologies for different applications. This cooperation can be undertaken within the framework of a public-private partnership such as America Makes.[2]**

## THE CHALLENGES OF SPACE-BASED ADDITIVE MANUFACTURING

Production of additive manufacturing components on the ground currently requires extensive human presence and participation. This is not always due to the complexity of the manufacturing operation; sometimes human labor to move parts from one machine to the next is cheaper than an automated system. Some automated manufacturing capabilities on the ground are currently under development, although it is not clear that a completely automated part-handling sequence of operations (e.g., setup, build, removal, finishing) is under development that would eliminate the need for human presence. Significant further development will be required for automated space-based additive manufacturing, and much of this development is likely to require government support. Spacecraft manufacturing is a conservative field, and private companies are reluctant to introduce advanced technologies on their own initiative. For this reason, government plays a vital role in conducting research that can ultimately benefit civil, military, and commercial satellite manufacturers.

Continuing development of terrestrial additive manufacturing processes will not necessarily drive investment or development of automation and human operator capabilities that might be translatable for space applications. Transplanting an additive manufacturing capability to space requires consideration of how the supporting infrastructure, including the applicability or desirability of maintaining humans in the loop, needs to be evolved to operate in the new environment. At the present time, the ISS offers an excellent research platform for additive manufacturing work. The ISS has the benefit of already being paid for. But the ISS will not exist forever, and in future decades, space-based additive manufacturing will require its own infrastructure support, with its own costs.

There are numerous potential benefits to using this technology in space. Additive manufacturing may provide

[2] America Makes is the current government-led consortium addressing additive manufacturing issues. NASA and AFRL both support America Makes.

entirely new capabilities. Both NASA and the Air Force have begun evaluations of small satellites, such as Cube-Sats, and NASA in particular is researching the role of space-based additive manufacturing for such satellites. NASA is currently evaluating the feasibility of this approach. The committee concluded that further evaluation of the costs and benefits of approaches to additive manufacturing in space should be conducted. In some cases, it may be possible to reduce costs by eliminating the requirement to launch spare or replacement parts into orbit.

**Recommendation: As the technology evolves and when projects utilizing this technology are considered, NASA and the Air Force should jointly undertake a cost-benefit analysis of the role of space-based additive manufacturing in the construction of smaller, more reliable, less massive satellite systems or their key components.**

By making baseline assumptions about scope and reference points, NASA and the Air Force can begin to define the overall parameters for additive manufacturing and its impact on spacecraft development. Such analysis, although limited in scope, can help to guide future decision making regarding research expenditures.

The infrastructure costs will have to be taken into account when evaluating the future potential of additive manufacturing in space, leading the committee to another recommendation:

**Recommendation: When considering moving additive manufacturing technology to the space environment, any person or organization developing plans should include in their planning the infrastructure required to enable fabrication processes based on additive-manufacturing, such as power, robotics, and even human presence. Studies examining the types of infrastructure should be undertaken in tandem with the development of the additive manufacturing technology itself.**

However, the committee was struck by the fact that additive manufacturing may also provide totally new capabilities. Thus, it would be a mistake to make additive manufacturing decisions based entirely upon cost-benefit determinations of existing products and functionalities, because doing so might lead to missing valuable opportunities to advance capabilities with this new technology.

**Recommendation: Actual costs of the reproduction of components or spacecraft should not be the sole criterion for evaluation of the benefits of additive manufacturing; criteria should also include the value of creating structures and functionalities not feasible before.**

## NASA AND ADDITIVE MANUFACTURING

Currently, NASA is the leader in space-based additive manufacturing. After first evaluating the technology in the late 1990s, the agency has sponsored an upcoming experiment aboard the ISS involving a small "3D printer" that will manufacture plastic parts that will be evaluated for quality and may be useful for operations. Many NASA field centers are currently conducting experiments with additive manufacturing on the ground, but only Marshall Space Flight Center is actively sponsoring space-based applications.

**Recommendation: NASA should consider additional investments in the education and training of both materials scientists with specific expertise in additive manufacturing and spacecraft designers and engineers with deep knowledge of the use and development of additively manufactured systems.**

The committee believes that this broad-based experimentation throughout the agency is valuable. However, it concludes that NASA will benefit from coordination of its many and diverse additive manufacturing activities. NASA's full use and application of additive manufacturing technologies, both in space and on the ground, could be made more efficient and effective with a stronger associative link between additive manufacturing technology and facility developers and users who may benefit in areas of efficiency, complexity, and cost reductions.

There are many impressive development efforts under way at companies and government-supported laboratories across the country, and NASA has already sponsored communications between interested groups in this area. Although much of this work is proprietary, it will be beneficial for NASA to learn about these developments and to encourage partnerships and sharing of ideas, leading the committee to the following recommendation:

**Recommendation: NASA should sponsor a space-based additive manufacturing workshop to bring together current experts in the field to share ideas and identify possible research projects in the short term (1-5 years) and medium term (5-10 years).**

NASA recently extended the lifetime of the ISS to 2024. The space station's lifetime could possibly be further extended. Nevertheless, this represents a finite opportunity for further development of the technology in an ideal environment, when human assistance is possible.

**Recommendation: NASA should quickly identify additive manufacturing experiments for all areas of International Space Station (ISS) utilization planning and identify any additive manufacturing experiments that it can develop and test aboard the ISS during its remaining 10 years of service and determine if they are worthy of flight. NASA currently has methods for providing research grant funding for basic research on additive manufacturing. The agency should closely evaluate funded research options to determine which would allow the most rapid transition of additive manufacturing to the ISS.**

Because of its broad-reaching activities involving additive manufacturing, NASA could consider creating an enduring forum devoted to additive manufacturing engineering technologies, focusing on serving all NASA centers, universities, small companies, and other organizations. Such a forum could function as a focusing element to orient the agency's efforts and activities in space-based and terrestrial additive manufacturing, providing a phased capability to identify, facilitate, integrate, and maximize attention and resources to this difficult, long-term objective of developing this technology for space use.

The committee also concluded that NASA needs to formally develop its plans for space-based additive manufacturing. Although the agency seems to be on a reasonable development path for this new but rapidly advancing technology, it is time for NASA to produce an agency-wide roadmap for space-based additive manufacturing.

**Recommendation: NASA should convene an agency-wide space-based additive manufacturing working group to define and validate an agency-level roadmap, with short- and longer-term goals for evaluating the possible advantages of additive manufacturing in space, and with implications for terrestrial additive manufacturing as well. The roadmap should take into consideration efficiencies in cost and risk management. NASA should build on the considerable experience gained from its Space Technology Roadmaps. The space-based additive manufacturing roadmap objectives should include, but not be limited to the following:**

- **Developing goals for using the technology to assist the agency in meeting its key missions, covering all appropriate mission directorates, especially long-duration human spaceflight and planetary operations, which would require defining, understanding, evaluating, and prioritizing the direct and supporting technologies for autonomously or minimally attended space-based additive manufacturing, and robotic precursor and free-flyer missions;**
- **Identifying flight opportunities, such as on the International Space Station, during its next decade of operations,**
- **Targeting the full technology-development life-cycle and insertion strategies through 2050, aligned with target agency missions, for all appropriate mission directorates, and related collaborations; and**

- **Ensuring that support for incremental advances to address the technical challenges is supplemented with support for activities related to reaching the full potential of additive manufacturing.**

Although 2050 is a long way into the future, NASA has recently announced long-range plans out to the mid-2030s, and the committee believes that technology development horizons should extend beyond current plans. Whatever date NASA decides on, it should be ambitious.

There is naturally some tension between the unfettered creativity and innovation inherent in new technologies and efforts to develop research plans and consensus built roadmaps, which may limit or discourage innovation. The committee believes that NASA is cognizant of this tension and can establish goals while still encouraging innovation. The previously mentioned workshop and forum are ways to accomplish this.

NASA plans to begin conducting experiments using a plastics-based 3D printer aboard the ISS starting in late 2014. In addition to further developing this technology, the next major steps will involve additive manufacturing using metals, which NASA is already evaluating and which the European Space Agency has also indicated it is researching. This technology poses many challenges for space use, including high power requirements. In addition, ground-based technology (such as the use of metal powders) may not be applicable to a microgravity environment. Nevertheless, this is an important technology, and developing a roadmap will help NASA clarify potential research paths.

NASA has ties with other agencies, including foreign partners on the ISS, and these contacts can provide benefits for further development of this technology.

**Recommendation: NASA should seek opportunities for cooperation and joint development with other organizations interested in space-based additive manufacturing, including the Air Force, the European Space Agency, the Japanese Space Agency, other foreign partners, and commercial firms.**

To prevent duplication of effort, the government-led consortium America Makes can serve a clearinghouse role by creating an additive manufacturing in space working group that includes participation from government, industry, academia, and international partners. Both NASA and the Air Force could be active leaders within the working group and ensure that each builds on the knowledge of the broader additive manufacturing community.

## THE AIR FORCE AND ADDITIVE MANUFACTURING

The committee found NASA's requirement for space-based additive manufacturing to be more clearly defined than the Air Force's requirements. The committee was informed that the Air Force's most pressing requirement is to reduce the cost of launching payloads to orbit. At the present time, it is too early to be certain that space-based additive manufacturing will make it possible to reduce the cost of space launch. It is also too early to determine how the Air Force may best make use of this technology, although its potential for the deployment of structures too large or fragile to fit in current launch vehicle payload shrouds could prove attractive for some national security missions.

There is at present a lack of knowledge to credibly determine whether or not development of an Air Force-specific space-based additive manufacturing production facility would achieve its expected benefit. Given that such a fabrication center would be highly complex and expensive, a detailed system assessment and cost-benefit analysis is advisable.

**Recommendation: The Air Force should conduct a systems-analytical study of the operational utility of spacecraft and spacecraft components produced in space using additive manufacturing compared to other existing production methods.**

Considering that the present state of manufacturing focuses on new types of individual components of specialized shapes, composition, and materials, it is clear that the task of manufacturing a complete scientific or military satellite of the complexity of current spacecraft is far in the future, if not impossible. This situation is unlikely to change unless major, very-long-term changes are made in this nation's space systems design, engineering methodologies, and infrastructure at all levels.

An independent, free-flying additive manufacturing satellite construction platform, human-tended or robotic, would require extensive ground-based development in additive manufacturing, robotics, and telepresence. Given the various limitations of power, cost, long build times, verification of manufacture, and other factors discussed previously, a large number of issues require resolution before committing to such a program.

**Recommendation: The Air Force should continue to invest in additive manufacturing technologies, with a specific focus on their applicability to existing and new space applications, and invest in selected flight experiments.**

The Department of Defense is already evaluating additive manufacturing technologies for a broad range of ground-based uses, including maintenance centers on Navy ships, Army field-based repair equipment, and design, fabrication, and repair of Air Force aircraft. This technology is sufficiently unique that it requires new skills and training, particularly for aerospace use.

**Recommendation: The Air Force should consider additional investments in the education and training of both materials scientists with specific expertise in additive manufacturing and spacecraft designers and engineers with deep knowledge of the use and development of additively manufactured systems.**

During its information gathering, the committee heard of relatively little Air Force involvement in planning for or developing additive manufacturing for space use. The committee concluded that the Air Force needs to start defining its requirements and research strategy for this technology in order to take advantage of and steer developments for its application to military space missions. The Air Force has well-developed and proven mechanisms for research planning.

**Recommendation: The Air Force should establish a roadmap with short- and longer-term goals for evaluating the possible advantages of additive manufacturing in space. The Air Force should build on the considerable experience gained from other Air Force technology development roadmaps. The space-based additive manufacturing roadmap should include, but not be limited to the following:**

- **Developing goals for using the technology in key Air Force missions, especially for autonomously or minimally attended, space-based additive manufacturing and free-flyer missions;**
- **Identifying flight opportunities, including those on non-Air Force platforms, such as the International Space Station, during its next decade of operations; and**
- **Targeting the full technology-development life-cycle and insertion strategies through 2050, aligned with Air Force missions, and related collaborations.**

Although the Air Force's path forward is not clear, the military can capitalize on the fact that NASA has already developed some of the infrastructure that will make it easier for the Air Force to research the potential capabilities of space-based additive manufacturing and is already engaged in current research of its own.[3] This provides an opportunity for the Air Force that it would not otherwise have.

**Recommendation: The Air Force should make every effort to cooperate with NASA on in-space additive manufacturing technology development, including conducting research on the International Space Station.**

---

[3] The committee sought, but was unable to find, historical data on U.S. government funding of additive manufacturing by various agencies.

If additive manufacturing in space does become commonplace, it could increase the debris generated in orbit. Both NASA and the Air Force are well aware of debris hazards and will undoubtedly include such considerations in their efforts.

## ORGANIZATION OF THIS REPORT

Chapter 1 introduces additive manufacturing and includes the committee's statement of task. Chapter 2 discusses potential uses and applications of additive manufacturing in space. Chapter 3 addresses the many technical and manufacturing issues that must be addressed, both terrestrially and extra-terrestrially, before creating and utilizing additive manufacturing in a space environment. Finally, Chapters 4 and 5 outline the next steps for NASA and the Air Force.

# 1

# Introduction

The Air Force and NASA have jointly asked the National Research Council (NRC) to conduct a study exploring the possibilities presented by new approaches to manufacturing space hardware, and especially to address the promise of additive manufacturing, specifically in space.

## THE POTENTIAL OF ADDITIVE MANUFACTURING IN SPACE

This report identifies the benefits, gaps between current and desired capabilities, and technology development paths for additive manufacturing's use in space systems. Although the report is aimed at in-space additive manufacturing and the benefits to space and non-space products it might bring, it will also make clear that space systems will have long-term benefits from and dependence on terrestrial additive manufacturing as well. The platforms the committee examined included ground-based test beds, the International Space Station (ISS), and human-tended platforms (both internal and external to the platform), free flyers (e.g., satellites), and non-terrestrial planetary-based platforms (e.g., in situ resource utilization and habitats).

In addressing its charge (Box 1.1), the committee explored the missions and space operating environments of the Air Force and NASA, assessing the applicability of additive manufacturing approaches, and identifying promising and potential results.

It is natural for NASA and the Air Force to explore opportunities for using additive manufacturing technologies in space where additive manufacturing could offer the potential to

- Reduce launch vehicle volumes as compared to an equivalent spacecraft and
- Enable tailoring of launch vehicle systems that deliver materials to orbit.

Both factors may contribute to improving launch economics. In addition, additive manufacturing in space could also

- Enable the design and manufacture of new materials and novel parts that have never been created before, potentially creating space-only parts that function well in zero gravity but not in a terrestrial environment;
- Transform operations and logistics planning via the ability to launch broad categories of materials that can be manufactured in situ into a range of parts with a wide variety of functionality; and

- Perhaps even transform the trade space when developing space hardware and robotic systems such that functional, small spacecraft can be fully manufactured in space to suit the needs of specific owners.

The committee has also identified a number of obstacles to achieving these desired outcomes, including a lack of clarity on mission scenarios that would drive the development of appropriate additive manufacturing technologies toward addressing real NASA and Air Force challenges, obsession with a new and novel technology without a clear eye toward potential costs, and lack of understanding of technical limitations and performance criteria.

To examine the possibilities for Air Force and NASA missions, the committee discussed the feasibility of the concept of space-based, additive manufacturing of space hardware (including, but not limited to, a fully functional

**BOX 1.1**
**Statement of Task**

The National Research Council will appoint an ad hoc committee to explore the implications of space-based additive manufacturing technologies for space operations and the manufacture of space hardware. In conducting the study and preparing its report the committee will:

- Assess the current state of additive manufacturing in the United States and worldwide (especially in the aerospace industries, universities, and national laboratories engaged in the design and manufacture of small satellites or respective subassemblies);
- Characterize the future states envisioned by the aerospace industries, universities, and national laboratories with respect to additive manufacturing and aerospace systems;
- Discuss the feasibility of the concept of space-based additive manufacturing of space hardware (including, but not limited to, a fully functional small spacecraft) that can conduct or enable missions of relevance to NASA, the Air Force, and/or the national security space communities;
- Identify the science and technology gaps between current additive manufacturing capabilities and the capabilities required to enable a space-based additive manufacturing concept, including those gaps that current trends indicate may be closed with commercial investments in additive manufacturing and those gaps that are likely to require dedicated investments by the federal government.
- Assess the implications that a space-based additive manufacturing capability would have on launch requirements (e.g., launching raw materials versus fully assembled spacecraft); overall satellite and payload designs; and in-space operations, such as possible reductions in mass and their implications for activities such as maneuverability.

The committee may also consider the following:

- The potential mission payloads and capabilities that could be expected from a space-based, additively manufactured spacecraft;
- The role in potential missions for a single spacecraft system manufactured in space by additive manufacturing or for multiple spacecraft systems, including disaggregated constellations and fractionated satellites;
- Concepts of operations for space-based manufacture of space hardware (including small spacecraft) using additive manufacturing, including development, test and evaluation, launch, deployment, and on-orbit command and control;
- Whether it is possible to develop a high-level heuristic tool that Air Force Space Command and other government organizations could use for first-order assessments of space-based, additively manufactured small spacecraft concepts in their integrated planning and process efforts.

small spacecraft). Where techniques are not yet mature or do not exist, the committee tried to identify the science and technology gaps between current additive manufacturing capabilities and the capabilities required to enable a space-based additive manufacturing concept.

The overall pace of implementation of additive manufacturing technologies will depend on the extent to which new engineering and testing protocols can be developed, evaluated and approved by professional organizations and the engineering and management communities in the aerospace industry and governments. New designs based on the unique materials, structures and manufacturing processes of additive manufacturing will need to prove their durability and safety for the applications in which they are targeted. Inclusion of additive manufacturing into all aspects of space operations may well extend over several decades, during which the various techniques for additive manufacturing hardware production are studied, tested, evaluated, and validated in a myriad of ways.

Additive manufacturing is already mature for a limited number of aircraft components and space-oriented components that could be manufactured on the ground. Yet the application of additive manufacturing in space is not feasible today, except for very limited and experimental purposes. Because of this, the Air Force and NASA will have to begin considering research and development (R&D) strategies to guide their investments wisely.

## DIFFERENT USERS, DIFFERENT REQUIREMENTS, OVERLAPPING TECHNOLOGIES

For NASA, high-quality work in space science and human exploration within an acceptable cost budget is most important. NASA's domains of interest include Earth, the Moon and other solar system objects, the disciplines of astrophysics and heliophysics, as well as human exploration of the surfaces of the Moon, asteroids and Mars. NASA is interested in additive manufacturing, seeking both to fulfill their responsibilities for advancing aeronautic and other technologies as well as finding cost effective ways to conduct scientific and exploration missions.

The Air Force has special responsibilities for understanding and taking advantage of additive manufacturing in the context of its responsibility to operate and sustain a fleet of approximately 55 spacecraft in five separate constellations, defined as follows:

- Protected communications (AEHF, Milstar);
- Wideband communications (WGS, DSCS);
- Missile warning (SBIRS, DSP);
- Position, navigation, and timing (GPS-III, GPS-II);
- Space situational awareness (SBSS, GSSAP); and
- Weather information (DMSP, future systems).[1]

In addition, a core Air Force mission is space superiority, and the Air Force operates many space- and ground-based systems to accomplish that mission. It is also actively developing new systems. Cost and speed of innovation are critical to maintain competitive advantage over potential adversaries, and additive manufacturing may be a critical technology to do that.

The Air Force Research Laboratory's (AFRL's) interest in in situ maintenance, repair, and production of Earth-orbiting space systems is a logical consideration related to reducing the costs of building and launching spacecraft built in ground-based additive manufacturing facilities; learning if additive manufacturing in space provides an effective means to lower cost satellites built in space; and determining if additive manufacturing can provide a means for space-based maintenance and repair to extend the lifetimes of satellites once they are in use.

With respect to reducing costs, aerospace companies are pursuing projects aimed at better understanding the value of additive manufacturing as a way to lower the costs of tooling and as a production tool for manufacturing key components of aircraft and spacecraft. Examples include key structural elements of high-performance fighter

[1] AEHF, Advanced Extremely High Frequency satellite; WGS, World Global Satcom; DSCS, Defense Satellite Communications System; SBIRS, Space-Based Infrared System; DSP, Defense Support Program; GPS, Global Positioning System; SBSS, Space Based Space Surveillance; GSSAP, Geosynchronous Space Situational Awareness Program; DMSP, Defense Meteorological Support Program.

aircraft and rocket engine components. Some companies are actively pursuing new means of constructing batteries and electric power and communication wiring to be integrated with satellite structures.

Finally, because of the Air Force's goal of seeking total system cost reduction, AFRL seeks advice on whether it may be possible to develop a means of using a build in space capability to undertake in situ satellite repair and/or maintenance involving bringing satellites back from their operational locations to a space-based repair and maintenance facility for repairs, upgrades, refueling, etc. This concept could result in a significant impact on the annual costs of the five constellations, depending on amortization of the costs involved, to create an additive manufacturing facility in low Earth orbit as well as its annual operating expenses, taken in the context of current program expenditures.[2]

## A RECENT HISTORY OF ADDITIVE MANUFACTURING

Additive manufacturing—commonly referred to as "3D printing"—is a general term encompassing various manufacturing methodologies, using different constructive materials and additive processes, each of which has specific advantages and constraints. In 2009, ASTM International formed the ASTM committee F42 on Additive Manufacturing Technologies to develop standards for additive manufacturing.[3] An important contribution of the ASTM F42 committee to date is a terminology standard that defines the different processes used to build three-dimensional parts from computer-aided design (CAD) files. Within this standard, there is a definition for "additive manufacturing," which is "a process of joining materials to make objects from 3D model data, usually layer upon layer, as opposed to subtractive manufacturing methodologies."[4] The definition adopted by ASTM International is the definition used in this report.

Additive manufacturing technology dates back to the 1980s when its industrial applications were largely seen as rapid prototyping of newly designed parts. This manufacturing method is currently receiving broad attention in U.S. industry and research universities, in part as a result of the recent creation of America Makes, formerly known as the National Additive Manufacturing Innovation Institute.[5]

America Makes is the current government-led consortium addressing additive manufacturing issues. NASA and AFRL both support America Makes, and it can serve an important role coordinating actions and avoiding duplication of effort, possibly with the creation of a working group on in-space additive manufacturing. Nevertheless, NASA has its own missions and interests and will have to take the lead to advance these.

Similar developments are also taking place within the European Union (EU). The EU agenda began in 2010 with the adoption of the Additive Manufacturing (AM) Sub-Platform 2010.[6] This work acted as a focal point for additive manufacturing development studies and resulted in the EU Strategic Research Agenda of 2013.[7] The overarching purpose is to provide guidance and coordination across the entire EU, where additive manufacturing is seen as one of the key enablers for long-term European economic progress, including the aerospace industrial sector. Organizations now involved with additive manufacturing studies include Germany's Fraunhofer Additive Manufacturing Alliance, focused on materials, technology, engineering, and quality; Universität Paderborn's Direct Manufacturing Research Center in Nixdorf, Germany; Belgium's Additive Manufacturing.be Network; and the multi-national European Additive Manufacturing Group, among many others.[8]

The European Space Agency (ESA) is actively exploring the role of additive manufacturing in space. In 2012, ESA conducted a study "Universal parts Fabricator-Replicator for Space Applications" that focused on using both

---

[2] The Defense Advanced Research Projects Agency (DARPA) conducted in-space servicing experiments with its Orbital Express spacecraft in 2007 and is currently sponsoring the Phoenix program which has analogous goals. Phoenix is discussed later in the report.

[3] ASTM International, formerly known as the American Society for Testing and Materials, is a globally recognized leader in the development and delivery of international voluntary consensus standards.

[4] ASTM F2792-12A, ref.

[5] See http://www.americamakes.us and http://www.ewi.org/additive-manufacturing-consortium/.

[6] See AM Platform website at http://www.rm-platform.com/.

[7] AM Sub-Platform, *2013 Additive Manufacturing: Strategic Research Agenda,* Version 2, http://www.rm-platform.com/linkdoc/AM_SRA_FINAL-V2.pdf.

[8] See European Powder Metallurgy Association, "European Additive Manufacturing Group (EAMG)," http://www.epma.com/european-additive-manufacturing-group, accessed March 11, 2014.

polymeric and metallic materials to develop replacement parts on the ISS. The Italian Space Agency (ASI) is funding a project to place a fused-deposition modeling (FDM) machine on the ISS.

The EU's AMAZE (Additive Manufacturing Aiming Towards Zero Waste and Efficient Production of High-Tech Metal Products) project involves 28 industrial partners across Europe and includes in-space applications as a core area. ESA is also supporting research on in situ additive manufacturing of habitats on the lunar surface using technology similar to that being developed in the United States. These activities indicate that there is an opportunity for international cooperation for developing this technology.[9]

There has also been extensive exposure of additive manufacturing in the public media of North America and Europe and throughout the globe, where the use of this technology is heralded as the beginning of a new era in consumer and engineering manufacturing. Much of this attention is related to the technology intersections of low-cost computer-based computational capability; advances in ease of use of CAD software; low-cost, high-precision XYZ platforms and controllers; and a variety of additive manufacturing techniques. Technological progress in methods and materials supported by research in industry, government laboratories, and academia is being exported to a new consumer base, and excitement for the technology is rising as low-cost systems enable a more diverse end-user community to acquire the technology for personal and commercial uses. Additive manufacturing is also driving new developments in materials science, manufacturing technology, and perhaps most importantly, substantial changes in the creative design/development process for end users.

Since the introduction of the first working 3D printer in 1984 by Charles Hull of 3D Systems, additive manufacturing has become increasingly important for traditional, ground-based production of consumer and industrial products. The most comprehensive source on the state of the additive manufacturing industry and technology is contained in the annual "Wohlers Report" produced by Wohlers Associates.[10] According to the *Wohlers Report 2014*, to date, 63 companies worldwide have manufactured more than 66,000 professional-grade additive manufacturing systems for eight principal industrial sectors. Of these eight sectors, the largest, at 21.8 percent, is consumer products and electronics, followed by motor vehicles at 18.6 percent, medical and dental uses at 16.4 percent, and industrial and business machines at 13.4 percent. Aerospace follows at 10.2 percent.

The dominant industrial user areas are determined by Wohlers[11] to be production of consumer products/electronics, followed by fabrication of parts for motor vehicles and machine parts for industrial and business equipment. The specific, dominant application areas for the products of these machines are given in Figure 1.1.

The production of functional parts for preproduction manufacturing activities, including presentation models, production of tooling components, patterns for metal castings, and patterns for prototype tooling, is 54 percent of the total use functions, and the creation of saleable, functional parts is 28 percent. The remaining 18 percent of uses are spread across four other areas. As these uses evolved and experience was gained in different industries, additive manufacturing products transitioned from industrial prototypes to applied parts used in engineering applications, thus driving the development of new and broader engineering standards for the additive manufacturing industry as a whole. Currently, terrestrial use of additive manufacturing in production has demonstrated some cost benefits for manufacturing very complex, end-item parts that do not lend themselves to linear or angular machining. Additive manufacturing is much too slow and expensive to compete with high-volume fabrication techniques such as injection molding, stamping, casting, or pressing when parts lend themselves to these conventional manufacturing processes.

According to recent industry reports, the sale of additive manufacturing machines for metal manufacturing in 2013 increased 76 percent over the previous year, and overall, the market for 3D-printing products and services grew to more than $3 billion in 2013, representing a growth of 35 percent over 2012. The primary buyers are in

[9] See Tommaso Ghidini, European Space Agency, "An Overview of Current AM Activities at the European Space Agency," presentation to the 3D Printing and Additive Manufacturing-Industrial Applications Global Summit 2013, November 19-20, 2013, http://www.3d-printing-additive-manufacturing.com/media/downloads/52-d1-12-20-c-tommaso-ghidini-esa.pdf, and European Space Agency, "3D Printing for Space: The Additive Revolution," October 16, 2013, http://www.esa.int/Our_Activities/Human_Spaceflight/Research/3D_printing_for_space_the_additive_revolution.

[10] T.T. Wohlers, *Wohlers Report 2014, 3D Printing and Additive Manufacturing State of the Industry*, Annual Worldwide Progress Report, Wohlers Associates, Inc., Fort Collins, Colo., 2014.

[11] Ibid.

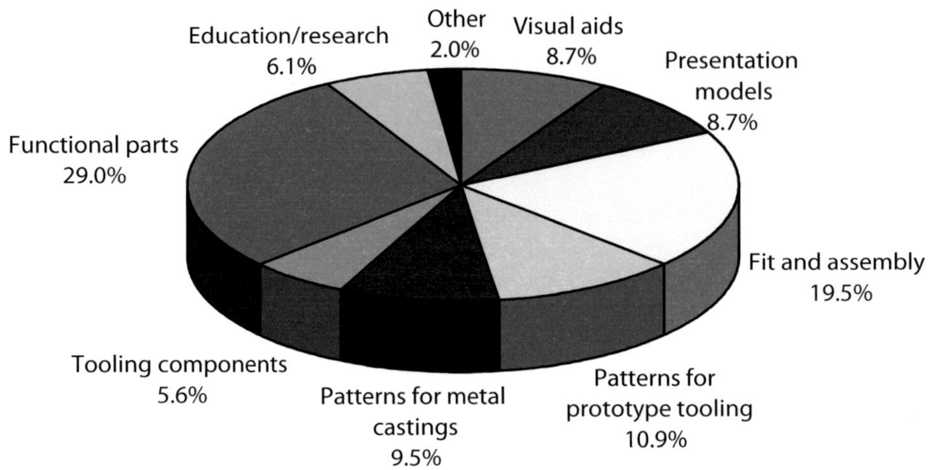


FIGURE 1.1 Uses of additive manufacturing systems, 2013. SOURCE: T.T. Wohlers, *Wohlers Report 2014, 3D Printing and Additive Manufacturing State of the Industry*, Annual Worldwide Progress Report, Wohlers Associates, Inc., Fort Collins, Colo., 2014. Courtesy of Wohlers Associates, Inc.

the medical (dental in particular) and aerospace industries, and with both prototyping and manufacturing uses. Aerospace companies especially are using 3D printers for testing and certification as they gear up for larger-scale manufacturing.[12]

Over the last quarter century, the United States led all other countries by a large margin (38 percent) of total industrial additive manufacturing systems installed. In the past 10 years, the United States led the world with 10 additive manufacturing companies, followed by Europe and Japan with 7 each, and China with 3. In more recent years, however, the geography of manufacturing and sales of additive manufacturing units has expanded and shifted. As of May 2013, 16 companies in Europe, 7 in China, 4 in the United States, and 2 in Japan produced and marketed additive manufacturing systems. Indeed, most companies that manufacture metal powder-bed fusion additive manufacturing systems are currently located outside the United States: 7 are in Europe, and 2 are in China.[13]

The total revenue from sales of additive manufacturing products and services (shown in millions of dollars) has been rising steadily since 1993 (Figure 1.2).

Market dynamics has successfully built the economic strength of additive manufacturing, supported by rapid commercial applications, company mergers and acquisitions, and investments via government interests. (A current perspective on the additive manufacturing industry and associated technologies can be found in the *Wohlers Report 2014*.[14]) A thorough discussion of the historical development of additive manufacturing technologies is provided by Gibson, et al.[15]

[12] See Wohlers Associates, Inc., "Metal Additive Manufacturing Grows by Nearly 76% According to Wohlers Report 2014," media release, May 21, 2014, http://wohlersassociates.com/press64.html, and Alex Knapp, "Sales of 3D Metal Printers Grew Over 75% in 2013," *Forbes.com*, May 21, 2014, http://www.forbes.com/sites/alexknapp/2014/05/21/sales-of-3d-metal-printers-grew-over-75-in-2013/.

[13] T. Wohlers, Tracking global growth in industrial-scale additive manufacturing, *3D Printing and Additive Manufacturing* 1(1):2-3, 2014, doi:10.1089/3dp.2013.0004.

[14] T.T. Wohlers, *Wohlers Report 2014*, 2014.

[15] I. Gibson, D.W. Rosen, and B. Stucker, Chapter 2, "Development of Additive Manufacturing Technology," in *Additive Manufacturing Technologies,* Springer Science+Business Media, 2010, doi:10.1007/978-1-4419-1120-9_2.

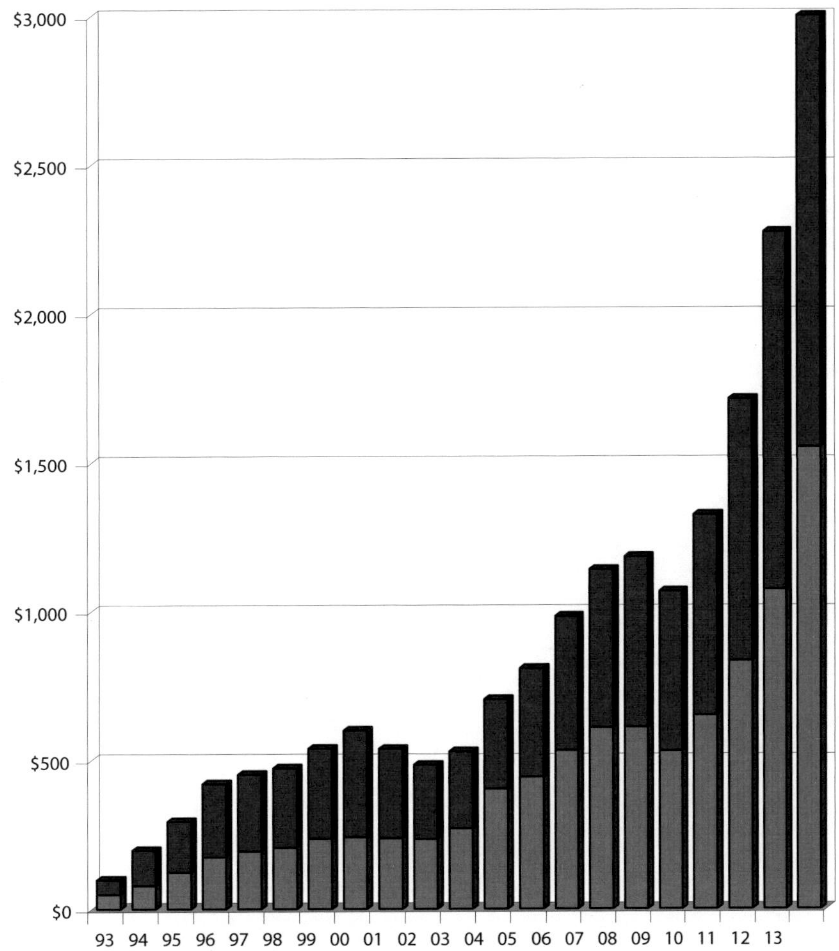


FIGURE 1.2 Global additive manufacturing revenues from products and services, 1993-2012 (vertical axis indicates millions of dollars and horizontal axis indicates years). Service revenues total $1.2 billion in 2012. The blue at the bottom indicates revenue from products sold, the red at the top indicates revenue from services sold. SOURCE: T.T. Wohlers, *Wohlers Report 2014, 3D Printing and Additive Manufacturing State of the Industry*, Annual Worldwide Progress Report, Wohlers Associates, Inc., Fort Collins, Colo., 2014. Courtesy of Wohlers Associates, Inc.

## STANDARDS FOR ADDITIVE MANUFACTURING

ASTM International is a global entity involved in the development and publication of international, voluntary, consensus technical standards. More than 12,000 ASTM standards have been developed for enhanced product safety, quality, market access, trade, and consumer confidence. In addition, a separate organization, the International Organization Standardization (ISO), is involved in technical product standards. Hence, standards for additive manufacturing are now being jointly developed by Committee F42 of the ASTM[16] and Technical Committee 261 of ISO through a first-ever agreement of its kind.[17]

The complexity of developing standards lies in the areas of terminology, processes and materials, test methods,

[16] See ASTM International, "Committee F42 on Additive Manufacturing Technologies," http://www.astm.org/COMMITTEE/F42.htm, accessed March 11, 2014.

[17] See Joint Plan for Additive Manufacturing Standards Development, ISO/TC 261 and ASTMF42, AM Standards Development Plan at http://www.astm.org/COMMITTEE/F42.htm.

and design and data formats. These in turn devolve into six different areas related to raw materials, processes and equipment, and finished parts. Figure 1.3 shows the standards structure for additive manufacturing.

To illustrate the complexity of what lies ahead for additive manufacturing, the following is a list of the high-priority standards that are current work items for ASTM/ISO and in-process for standards development:

- Qualification and certification methods,
- Design guidelines,
- Test methods for characteristics of raw materials,
- Test methods for mechanical properties of finished additive manufacturing parts,
- Material recycling (reuse) guidelines,
- Standard protocols for round-robin testing,
- Standard test artifacts, and
- Requirements for purchased additive manufacturing parts.

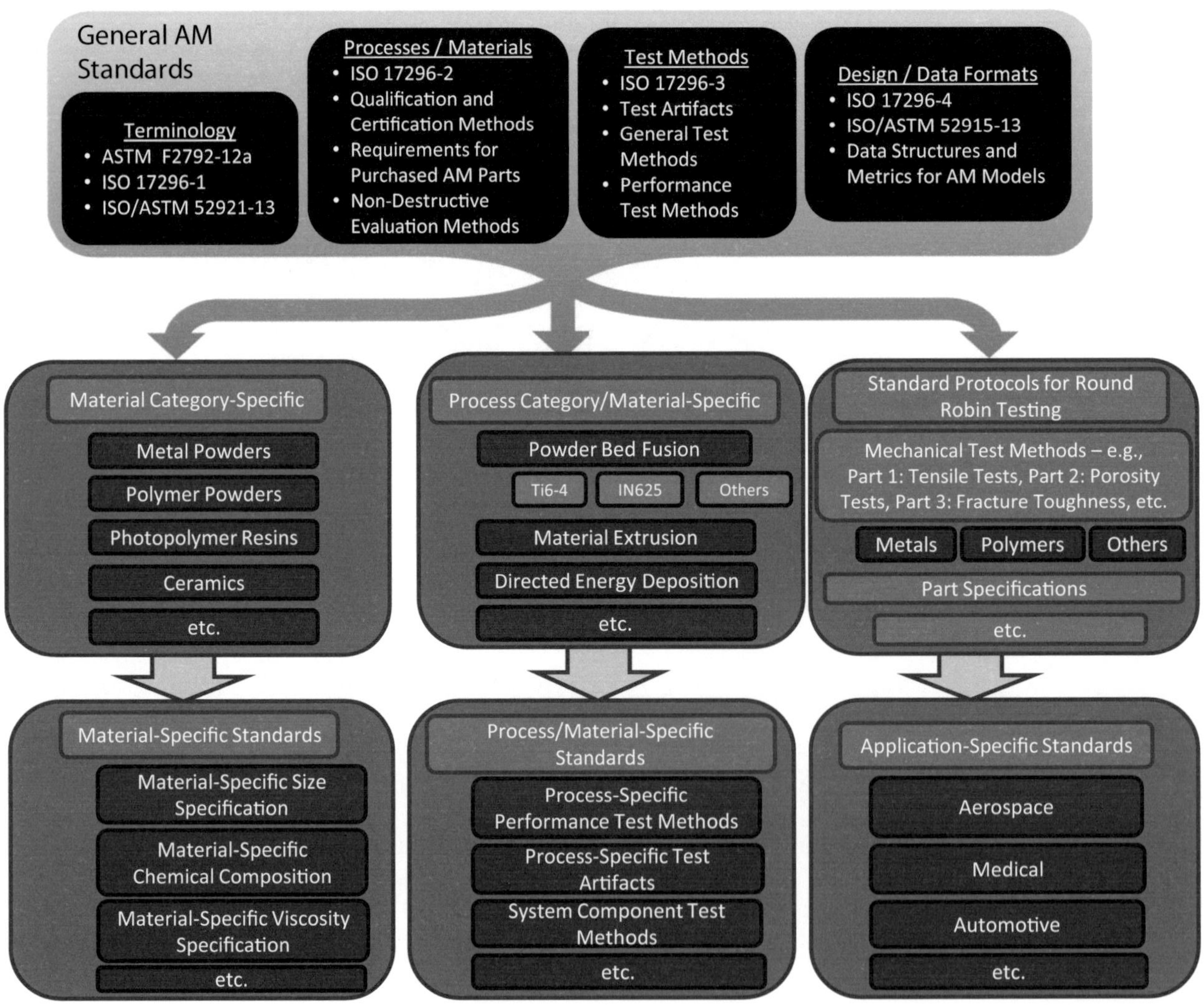


FIGURE 1.3 Structure of additive manufacturing (AM) standards for ASTM and ISO. SOURCE: Courtesy of ASTM Committee F42 on Additive Manufacturing Technologies, copyright ASTM International.

Broad-scale use of additive manufacturing within the fields of engineering and product development will not occur until these standards have been accepted by ASTM and ISO and accepted by the tens of thousands of manufacturing organizations worldwide. It is on this basis that additive manufacturing, as it exists today, is not a widely and fully accepted manufacturing process.

## HARMONIZATION OF EXISTING TERMINOLOGY STANDARDS

The ASTM currently recognizes specific terminology for fundamental additive manufacturing. The seven categories given for additive manufacturing technologies under ASTM Standard F2792-12A are outlined below.

### Vat Photopolymerization

In vat photopolymerization, a liquid photopolymer contained in a vat is selectively photocured using an energy source. Common vat photopolymerization technology includes a laser that scans a beam across a vat of photopolymer and another that projects the entire area image (area processing) onto the liquid surface using a light source and an image projection system. Common processes in this category include the first commercialized technology, stereolithography, known as SL or SLA (for stereolithography apparatus), and DLP for the area projection technology (because the area projection technology employs a Texas Instruments Digital Light Processing chip). There are numerous trademarked terms used in the additive manufacturing industry by equipment manufacturers for particular machines. It is not the intent of this report to provide an exhaustive list of these specific terms, although some are highlighted for reference.[18]

### Material Extrusion

In material extrusion, 3D parts are constructed layer by layer using materials extruded through a nozzle or orifice that is placed in desired regions using some form of translation mechanism. One implementation of this technology selectively deposits thermoplastic material through a heated nozzle, much like a glue gun placed on stages that move the nozzle to selectively deposit the material. FDM is an example of a process in this category, which is widely used in prototyping shops.

### Material Jetting

Material jetting selectively deposits droplets of material onto a platform. In common implementations, photopolymer and wax-like materials are used. Multi-jet modeling (MJM) and PolyJet are two commonly used names referring to particular machines that use material jetting. These systems typically use print heads with multiple nozzles that are capable of printing parts with multiple materials.

### Binder Jetting

Binder jetting is similar to material jetting except the build material is not selectively deposited. Instead, a binder or glue material is selectively deposited onto a bed of particles contained in some type of container or vat. The particles are glued together in the regions where the binder is deposited. After binder deposition, a platform is moved downward a distance equal to one layer of thickness, and a new layer of particles is raked over the build container from a powder source. There can be more than one source of powder providing the new powder to the build container. This process was originally developed at the Massachusetts Institute of Technology, which was called 3D printing (but never trademarked), and licensed to several companies that continue to provide machines today.

[18] For more complete lists, the interested reader can refer to, for example, T.T. Wohlers, *Wohlers Report 2014*, 2014.

### Powder Bed Fusion

Powder bed fusion represents a group of technologies that typically use either polymer or metal powders contained in a build container or vat. The material is selectively bound together using a scanning energy source, typically a laser or an electron beam. Parts are built in a build chamber on a platform that moves downward after each layer is fabricated. As each layer is completed, a new powder layer is raked to provide a thin and uniform layer of new powder over the previously fabricated layers. Common terminology for processes that employ this technique include selective laser sintering, laser sintering, selective laser melting, direct metal laser sintering, electron beam melting, and others.

### Sheet Lamination

Sheet lamination printing technologies bond sheets of material together to form the 3D shapes. In the additive process, a sheet of material is bonded on top of a previous sheet using glue, ultrasonic consolidation, or some other method. Typically, sheet lamination processes require combining additive manufacturing with some form of subtractive manufacturing process such as cutting or machining. After depositing a sheet (or possibly multiple sheets), a cutting or milling mechanism is used to define the features of the layer(s). In the case of machining, an end mill can machine away unwanted material for that layer, or in another case, a knife can be used to cut out the desired features for the current layer. Subsequent layers are deposited, and the subtractive process is repeated, as required, for each layer. The first commercialized technology in this category was referred to as Laminated Object Manufacturing, although this company is no longer in business. Today, there is a paper-based technology on the market as well as one that uses thin metal tapes bonded together using a process called ultrasonic additive manufacturing.

### Directed-Energy Deposition

In directed-energy deposition, three-dimensional shapes are constructed using lasers or electron beams directed at the build surface, with material fed into the build region to coincide with the incident energy source. A wire feed or powder feed system is used to deliver material into the build zone. Two common processes in this category include laser-engineered net shaping, which uses a laser with a powder feed system that enables more than one material to be deposited simultaneously, as well as direct manufacturing, which uses an electron beam and a wire feed system. Each of these processes have benefits and trade-offs when compared to one another, including material choice, build speed, layer thickness, surface quality, cost, and feasible part geometries, among others.

As for terrestrial additive manufacturing and manufacturing in general, issues like intellectual property, cybersecurity, counterfeit parts, and so on will have to be addressed by the community. Those working on space-based additive manufacturing will have to determine if there are any additional considerations unique to their field.

## GROUND-BASED ADDITIVE MANUFACTURING FOR AEROSPACE USE

Additive manufacturing applications are advancing significantly for aerospace uses. Lockheed Martin used additively manufactured brackets for microwave communication parts for NASA's Juno spacecraft launched in 2011, which is now under way toward Jupiter (Figures 1.4 and 1.5).

Other aerospace companies are exploring the use of additive manufacturing in spacecraft products or development projects in an effort to reduce time and costs. For example, Aerojet Rocketdyne recently used additive manufacturing to manufacture and successfully test a LOX/H2 rocket engine injector.[19] The same company is also

[19] A video clip of the test firing of this rocket engine is at http://www.youtube.com/watch?feature=player_embedded&v=40R9GQjawTE.

FIGURE 1.4 The Juno spacecraft, which will orbit Jupiter in 2016, includes the first known additively manufactured space system component, made by Lockheed Martin. SOURCE: Courtesy of NASA/JPL.

FIGURE 1.5 Additively manufactured waveguide brackets (shown by red arrows) installed on the Jupiter Juno spacecraft during assembly. SOURCE: Courtesy of Lockheed Martin.

FIGURE 1.6 Examples of small aircraft produced partially or entirely with additively manufactured parts by Aurora Flight Sciences. The increasing sophistication of machines produced with additive manufacturing technology is relevant to the in-space production of parts and even entire small spacecraft. SOURCE: Courtesy of Aurora Flight Sciences.

offering for sale four different space-qualified thruster systems produced with additive manufacturing for CubeSats and other small satellites.[20]

Other aerospace groups focusing on smaller constructed artifacts have made significant advances in using additive manufacturing for their products. In 2012 Aurora Flight Sciences of Manassas, Virginia, built and flew a thermoplastic drone system constructed via FDM using a commercial 3D printer. Aurora has produced and flown several 3D-printed aircraft to date. These aircraft, shown in Figure 1.6, have been used to prove-out aerodynamic designs before going into higher-volume production, as well as innovative structural arrangements not possible using traditional construction methods. Aurora has also flown aircraft with active sensors embedded into 3D-printed wings and printed heat exchangers using novel materials in the 3D-printing process. Similarly, the European company Airbus Group (formerly EADS), in collaboration with faculty and students at Leeds University, England, recently constructed a fully functional metallic prototype of the airframe and propulsion system for a drone aircraft. The prototype was created with a commercial additive manufacturing machine in a university laboratory, and a full flight-capable version is anticipated. In addition, the University of Sheffield is conducting similar work, and others are rumored to be investigating additively manufactured small unmanned aerial systems as well. These examples demonstrate that additive manufacturing is able to create complex geometrical shapes, many of which are impossible to produce with traditional casting or machining.

[20] D. Schmuland, C. Carpenter, R. Masse, and J. Overly, "New Insights into Additive Manufacturing Processes: Enabling Low-Cost, High-Impulse Propulsion Systems," 27th Annual AIAA/USU Conference on Small Satellites, AIAA Paper SSC13-VII-4, 2013, American Institute of Aeronautics and Astronautics, Reston, Va.

Another area where additive manufacturing holds intriguing promise is in the potential to create materials, and combinations of materials, in support of creating specific product material properties. A group at the Jet Propulsion Laboratory (JPL) is advancing the creation of objects with custom compositional gradients within metal structures, allowing engineers to design additively manufactured objects with localized, specific values of selected physical characteristics, such as an artifact's hardness, rigidity, and/or electrical and thermal conductivities. These manufacturing opportunities present the possibility of previously unattainable material combinations.[21] New technically sophisticated parts tailored for performance under various structural load and temperature conditions are emerging. An example of this new capability, a gradient-metallic alloy mirror assembly developed at JPL, is shown in Figure 1.7.

These unique mechanical parts are illustrative of the extraordinary opportunities available to enhance capabilities of equipment exposed to the harsh and complex physical conditions of space.

In the 2000s, the Defense Advanced Research Projects Agency launched its Direct Write program to develop the means to fabricate various types of conformal integrated electronic components, such as power supplies, connectors, application-specific integrated circuits, and many different types of environmental sensors. This initiative and its follow-on activities have contributed to integrating electric and electronic devices and systems into additively manufactured structures.

It may soon be possible to produce inexpensive electronic circuits imbedded in or on the surface of larger structures using additive manufacturing or direct-write machines. Figure 1.8 illustrates recent printed conductor work of Simon Leigh and colleagues at Imperial College, London, under the aegis of the United Kingdom's Research Centre in Nondestructive Evaluation.[22]

Finally, additive manufacturing offers economic incentives during the manufacturing process as compared to typical (subtractive) manufacturing processes. From a raw materials perspective, additive manufacturing creates a minimum of manufacturing byproduct (waste). From a bill of materials perspective, depending on the material type and form, cost savings of more than 75 percent can accrue from using additive manufacturing rather than milling methods of material removal. Current demonstrations of additive manufacturing production of simple part types show improvement in speed of product creation by about 40 percent. However, currently, the variety of materials (metals and plastics) available to support additive manufacturing is only a small subset of those used in subtractive manufacturing.

Research is under way to develop more sophisticated hybrid additive manufacturing systems that combine additive manufacturing machines with direct-write machines and other manufacturing technologies to enable embedding of electronic components and circuitry in three dimensions during fabrication. Figure 1.9 demonstrates several examples of structures with integrated electronics fabricated via experimental additive manufacturing systems.

Recent innovations in the additive manufacturing industry involve building machines capable of adding additive manufacturing materials to preconstructed metallic or other substrates. This enables a considerable savings in construction time for specific items where a casting or other manufactured item is used as a foundation for a more complex surface put in place via additive manufacturing. Other, new machines include combinations of additive manufacturing and subtractive tools to speed up the development of prototype tools or other constructed objects.

While the many benefits and potential of additive manufacturing for ground-based aerospace applications are clear from the preceding discussion, there are also currently various disadvantages that must be considered in the application of this technology. These are predominantly materials and processing related. Disadvantages of additive manufacturing have been well documented and include such issues as cost of operation (equipment, maintenance, and materials); machine performance (size, speed, reliability, repeatability, and reproducibility of the produced parts); availability of materials. These areas are currently topics of much research, and methods to overcome them will be available in time.

---

[21] D.C. Hofmann, J.P.C. Borgonia, R.P. Dillon, E.J. Suh, J.L. Mulder, and P.B. Gardner, "Applications for Gradient Metal Alloys Fabricated Using Additive Manufacturing," NASA Technical Brief, Jet Propulsion Laboratory, October 1, 2013, http://www.techbriefs.com/component/content/article/17446.

[22] S.J. Leigh, R.J. Bradley, C.P. Purssell, D.R. Billson, and D.A. Hutchins, A simple, lowcost conductive composite material for 3D printing of electronic sensors, *PLoS ONE* 7(11): e49365, 2012, doi:10.1371/journal.pone.0049365.

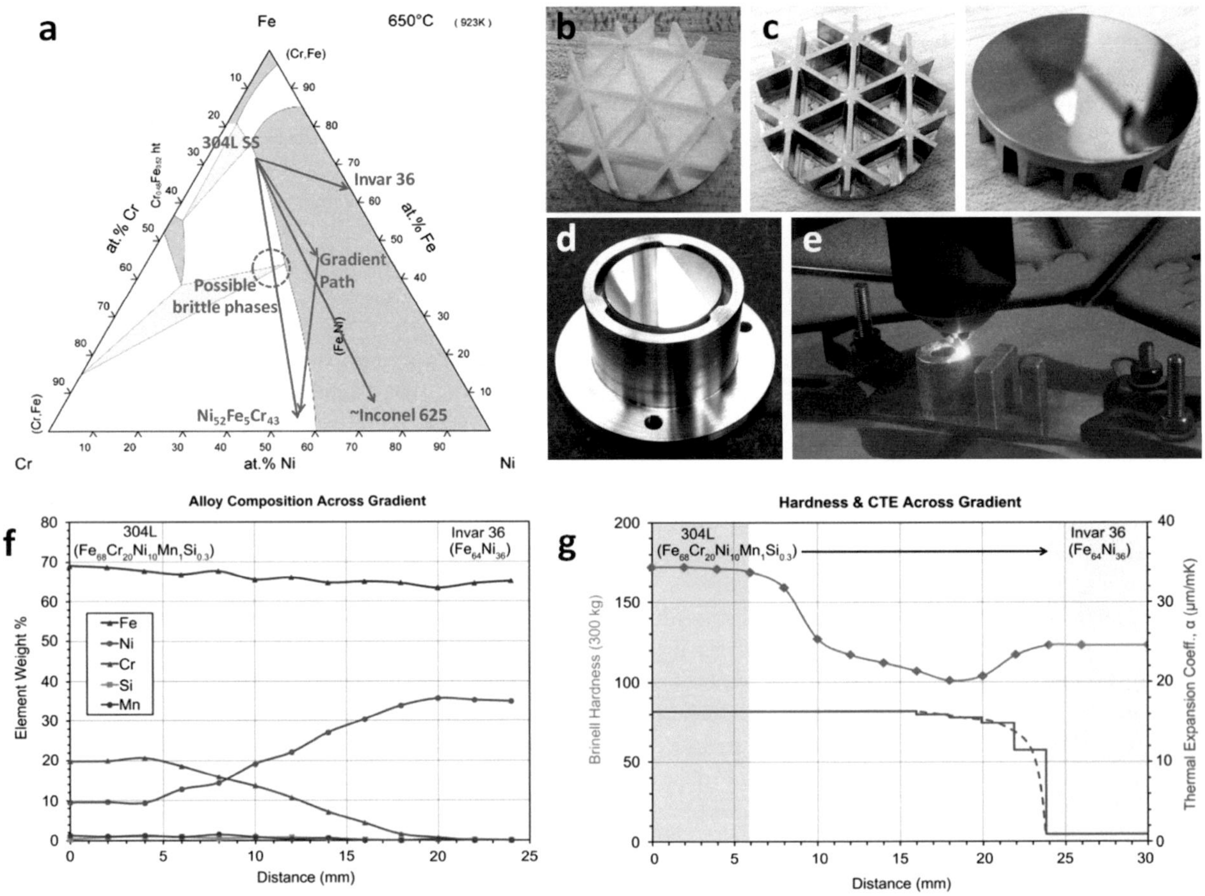


FIGURE 1.7 (a) An isothermal slice of the Fe-Ni-Cr ternary phase diagram showing how different gradient compositions can be mapped. The lines represent composition gradients between 304L stainless steel and Invar 36, a simplified Inconel 625 alloy and a NiFeCr alloy. In some gradients, the path intersects brittle intermetallic phases, which can be avoided by changing the path to go through more desirable phases (the segmented green line). (b) An isogrid mirror fabricated using a 3D plastic printer and (c) the same part fabricated using laser-engineered net shaping (LENS) after some finish machining. The mirror surface is made of Invar 36 and the isogrid backing is a gradient alloy that transitions from Invar 36 to stainless steel. (d) A gradient alloy mirror assembly with a metal-coated glass mirror attached to the Invar side of the assembly using epoxy. The mirror assembly transitions into stainless steel at the base. (e) Test samples of a Ti-V gradient alloy being fabricated by LENS. (f) The compositions (as measured through electron dispersive XRD) of the gradient mirror assembly in (d) showing the transition between Invar and stainless steel. (g) A plot of hardness and thermal expansion across the gradient mirror assembly. The intermediate phases of the gradient have been designed to be soft austenite (as demonstrated by the decreased hardness). The controllable thermal expansion makes this part alluring for optics applications. One side of the gradient has a near-zero thermal expansion, while the other side matches steel. SOURCE: NASA, Tech Briefs, "Applications for Gradient Metal Alloys Fabricated Using Additive Manufacturing: A New Roadmap for Gradient Metals that Could Be Used in Cars, Optics, Aircraft, and Sporting Goods," Jet Propulsion Laboratory, Pasadena, Calif., October 1, 2013, http://www.techbriefs.com/component/content/article/5-ntb/tech-briefs/materials/17446. Images courtesy of NASA/Jet Propulsion Laboratory/California Institute of Technology.

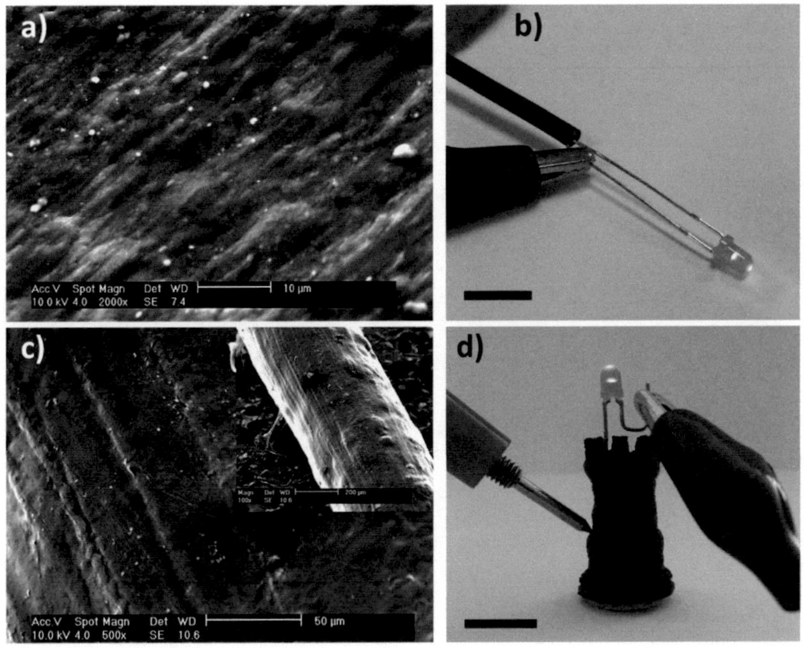


FIGURE 1.8 High-resolution scanning electron microscope images show both the cut edge of the conducting material (panel a) and the ink-jet printed, electrically conducting materials (panel c). Panel (b) shows a drawn length of the conductor being used to light a 4 mm light-emitting diode (scale bar 5 mm) and panel (d) shows a 3D printer-constructed chess piece being used as a conductive link (scale bar 10 mm). SOURCE: S.J. Leigh, R.J. Bradley, C.P. Purssell, D.R. Billson, and D.A. Hutchins, A simple, low-cost conductive composite material for 3D printing of electronic sensors, *PLoS ONE* 7(11): e49365, 2012, doi:10.1371/journal.pone.0049365.

## ADDITIVE MANUFACTURING CONSTRUCTION OF SPACECRAFT ON EARTH

At the present time, an active area of additive manufacturing for space applications is the development of CubeSats. First conceived and developed in the late 1990s by faculty and students at California State University, San Luis Obispo, and Stanford University, the first CubeSat was launched into low Earth orbit (LEO) in 2003 aboard a Russian rocket.[23] Today, more than 100 of these small (the basic unit [U] size is a base of 10 cm × 10 cm and a height of some multiple of 10 cm, upgradable to 60 cm) satellites have now been placed in LEO for the needs of a wide variety of technical applications. Figure 1.10 shows the increasing popularity of CubeSats among a range of private, government, and public organizations worldwide. Information about specific CubeSat missions, those already flown and in the queue for future flights, is available on the Internet. In addition to those flown previously, more than 90 CubeSats were launched during 2013, including the release of 28 platforms from the Minotaur I ORS-3 launch of November 2013 and 12 platforms from the NROL-39 Atlas V GEMSat launch of December 2011.[24] There was a significant increase in the number of commercial CubeSats launched in 2013 as well, indicating a transition from government and R&D activities to a much broader user community. Although additive manufacturing's benefits can accrue to any size satellite, the low cost and relative simplicity of CubeSats makes them ideal for experimentation with this new technology.

CubeSats were originally built using traditional spacecraft technologies. More recently, many are being built with a wide range of components and external structures produced with additive manufacturing materials. The

---

[23] Debra Werner, "Profile: Jordi Puig-Suari, Co-Founder of Tyvak Nano-Satellite Systems LLC," *Space News,* August 13, 2012.

[24] These CubeSats were provided through NASA's Educational Launch of Nanosatellites (ELaNa) program and the National Reconnaissance Office's Mission Integration Directorate.

University of
Texas at El Paso

A

B

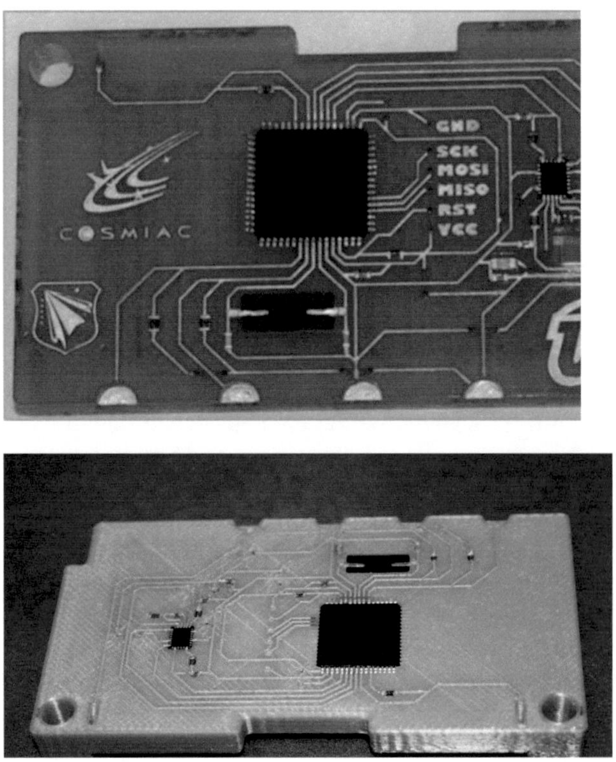
GND
SCK
MOSI
MISO
RST
VCC
COSMIAC

C

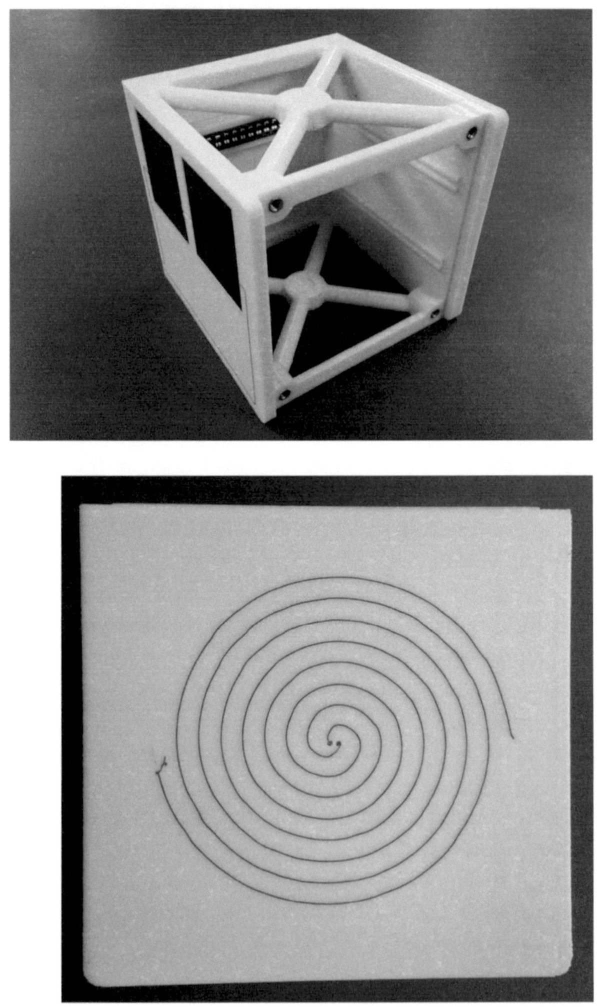

D

overarching goal is to simplify construction and mass of all on-board equipment, including power, communication, propulsion, thermal control, attitude control, digital systems, and instrumentation to the greatest extent possible.

One such CubeSat, Montana State University's (MSU's) PrintSat, is a good example of the advantages of additive manufacturing. This satellite was built to demonstrate the utility of using additive manufacturing for space structures and mechanisms. The satellite structure is printed from Windform XT 2.0, a polyamide-based carbon-filled material used for demanding terrestrial applications. The payload elements, which are not additively manufactured, will include a single-chip hybrid radiation micro dosimeter, a loads cell, and a surface resistivity sensor to measure the surface resistivity of the satellite's nickel plating. Laboratory measurements will be made on an identical structure to enable evaluation of changes expected while PrintSat is in orbit. Full telemetry with data, housekeeping, and Global Positioning System-determined attitude and location will be enabled. The overall system was designed by faculty and students of MSU's Space Science and Engineering Laboratory using CADCAM and other engineering tools. Implementation of the designs was done by CRP-USA at their U.S. production facility. Complete flight testing of PrintSat was successfully done to NASA standards at MSU, and the flight-ready package was sent to Sandia National Laboratories for launch aboard a small launch vehicle known as Super Strypi as part of Sandia National Laboratories' Operationally Responsive Space-4 program. Figure 1.11 gives views of PrintSat's key features.

## ADDITIVE MANUFACTURING CONSTRUCTION IN SPACE

NASA, the Air Force, other government agencies, universities, nonprofit organizations and commercial aerospace firms are now exploring various forms of additive manufacturing for designing, constructing, and operating individual components, subsystems, and entire systems for a wide range of autonomous space systems. In addition, additive manufacturing applications may influence and benefit human-related systems or facilities operated temporarily or permanently in space or, prospectively, on or near planetary bodies such as the moons of Earth or Mars, asteroids, or Mars. Such applications may emerge in the future and will likely benefit from the application of the new additive manufacturing processes suiting the rigorous environmental and functional constraints associated with space conditions. They will also emerge as professional confidence is gained in the use of new engineering standards.

## A BRIEF HISTORY OF SPACE-BASED CONSTRUCTION

Construction in space dates to the earliest days of the space program. Early concepts involved the integration of large, complex components such as space station modules, rather than the manufacturing of parts or components in space. For most of the history of the space program, space operations required either launching fully integrated spacecraft or connecting components in orbit either robotically or with human assistance. Even simple structures and objects were entirely manufactured on the ground and launched into space and connected by conventional methods. Although the Soviet space program conducted on-orbit welding experiments, and in-space welding has

FIGURE 1.9 Example structures with embedded electronics fabricated via additive manufacturing combined with direct-write and other manufacturing technologies. (A) Magnetic flux sensor system with curved surfaces and modern miniaturized electronic components (microcontroller, conductive ink interconnect, light-emitting diodes, Hall effect sensors, and power supply connector) printed via stereolithography and direct print technologies; (B) Novelty six-sided gaming die with microprocessor and accelerometer manufactured via stereolithography and direct print technologies; (C) CubeSat module produced by stereolithography and direct print technologies (*top*) as well as fused-deposition modeling (FDM), CNC routing, and direct print technologies (*bottom*); (D) CubeSat 1-unit (1u) housing produced by FDM incorporating solar cells and signal and power buses into a polycarbonate substrate (*top*) as well as a surface antenna embedded via ultrasonic wire embedding technology (*bottom*). SOURCE: (A, C, and D) W.M. Keck Center for 3D Innovation, University of Texas, El Paso; (B) E. MacDonald, R. Salas, D. Espalin, M. Perez, E. Aguilera, D. Muse, R.B. Wicker, 3D printing for the rapid prototyping of structural electronics, *IEEE Access* 2:234-242, 2014, doi:10.1109/ACCESS.2014.2311810.

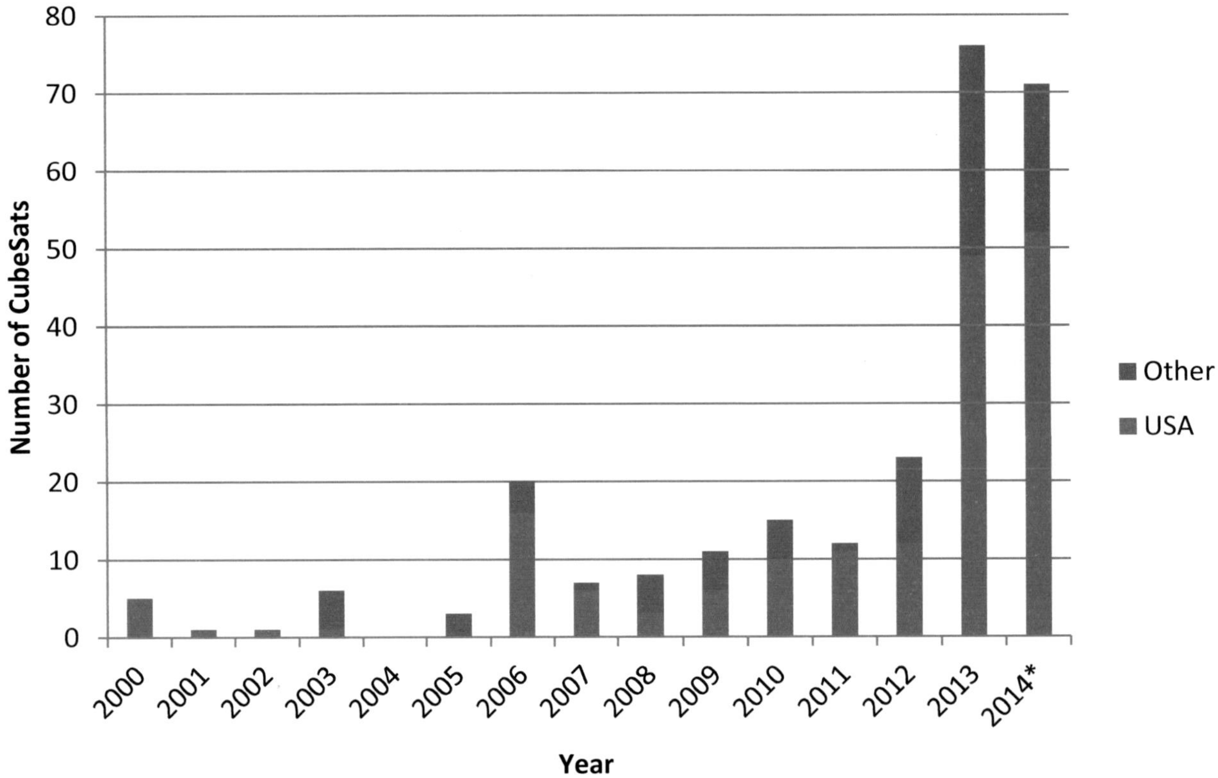


FIGURE 1.10 CubeSat launches by year, 2000 to July 2014. The number of CubeSats manifested per year are sorted by the sponsoring organization's country of origin. These numbers are higher than the number of operational CubeSats, because some suffered launch failures, deployment failures, and so on. NOTE: *, launches as of July 14, 2014. SOURCE: Data were compiled from M. Swartwout, The first one hundred CubeSats: A statistical look, *Journal of Small Satellites* 2(2):213-233, 2013, Appendix A (*2000-2012*) and G. Krebs, Gunter's Space Page, "CubeSat," http://space.skyrocket.de/doc_sat/cubesat.htm, accessed July 14, 2014 (*2013-2014*).

been studied by at least one U.S. aerospace contractor, neither NASA nor other space programs chose to employ supporting technologies, such as welding for space construction, and the ISS was largely an assembly and integration project with all manufacturing performed on the ground.

However, there have been several proposals for in-space manufacturing of construction materials. In the mid-1970s, under NASA contract, Grumman Aerospace built a Space Fabrication Demonstration System, also known as a "Beam Builder (B2)," that was capable of assembling triangular cross-section aluminum truss structures. This device was tested at NASA Marshall Space Flight Center (MSFC) (Figures 1.12 and 1.13).

In addition to the Beam Builder, NASA considered the possible use of in situ materials, such as lunar regolith or martian soil, for the construction of structures. But no substantial work has been done on this subject.

## A BRIEF HISTORY OF ADDITIVE MANUFACTURING ABOARD THE ISS

Prior to the development of the ISS, there was a detailed study and experimental testing of additive manufacturing for space applications. This was the work begun in 1999 by K.G. Cooper and M.R. Griffin, employees

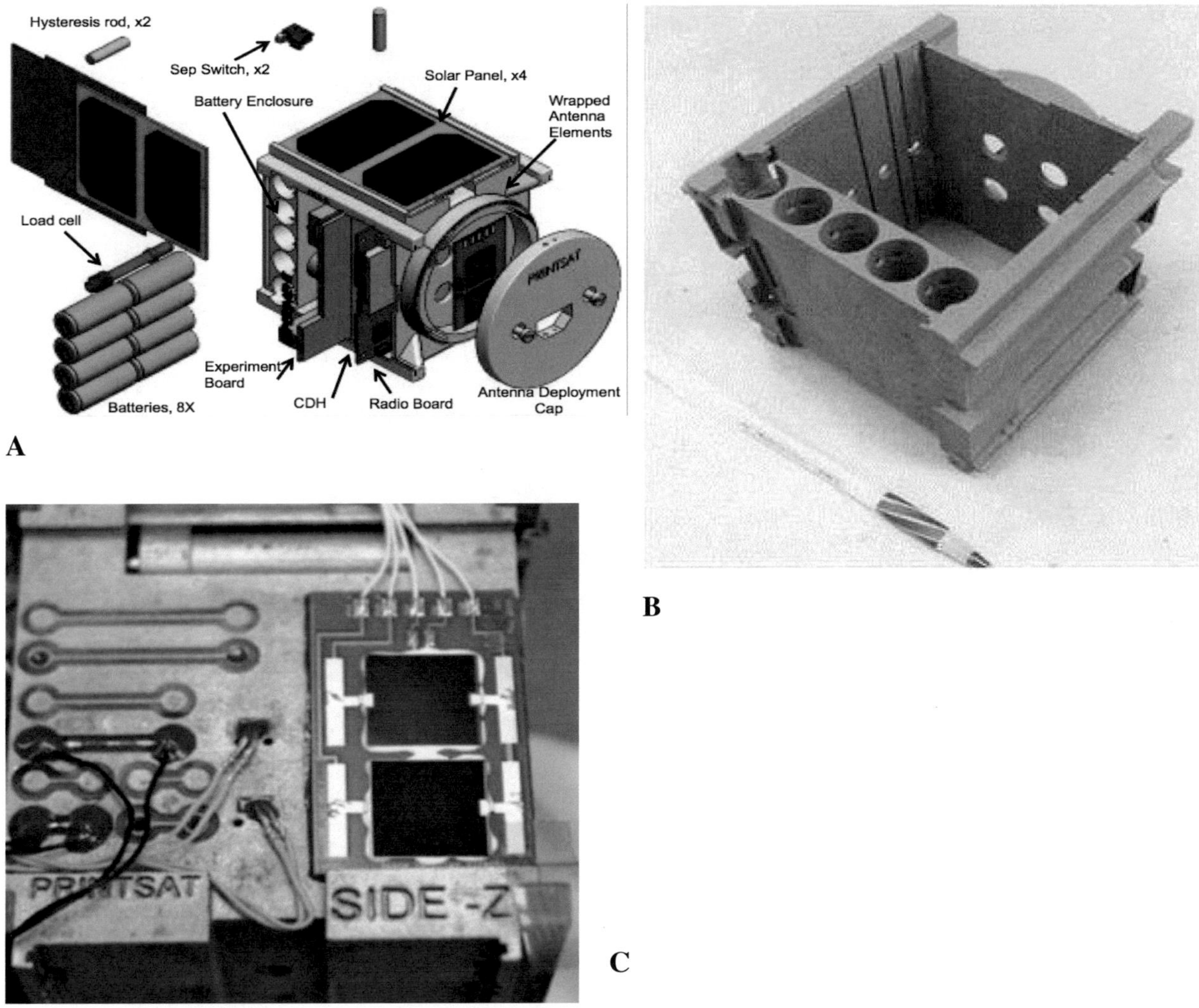


FIGURE 1.11 (A) The principal components of the PrintSat spacecraft, (B) overview of the PrintSat external structure, and (C) space environment effects surface sensors and test solar cells. PrintSat is scheduled for launch to low Earth orbit in the first quarter of 2015 aboard a Super Strypi rocket. PrintSat is the product of a consortium led by Montana State University and implemented in Windform XT 2.0 carbon-fiber-reinforced composite material. Additive manufacturing printing was done by CRP-USA (see http://www.crp-usa.net/). SOURCE: Courtesy of the Space Science and Engineering Laboratory at Montana State University on behalf of the PrintSat team.

of NASA MSFC.[25] In 2000, the NRC published a report that referred to "direct manufacturing" (another term sometimes used for additive manufacturing) and stated, "In remote locations such as the Moon or Mars, direct fabrication from computer numerical control programs could be used to produce items on location, reducing reliance on spare parts inventories."[26]

[25] K.G. Cooper and M.R. Griffin, *Microgravity Manufacturing Via Fused Deposition*, NASA/TM-2003-212636, Marshall Space Flight Center, Huntsville, Ala., July 2003.

[26] The report also noted that the concept of a "universal, compact machine shop with a very broad capability that might even extend to repairing itself would be included in the spacecraft or at the base" originated in 1987, but was now becoming conceivable because of additive manufacturing (NRC, *Microgravity Research in Support of Technologies for the Human Exploration and Development of Space and Planetary Bodies*, National Academy Press, Washington, D.C., 2000, pp. 99-100).

FIGURE 1.12 Grumman"Beam Builder" machine tested at Marshall Space Flight Center in the late 1970s. The machine used three rolls of rolled aluminum that it bent and then welded with cross braces (the cross braces were stored in the orange cartridges visible on the side). Grumman followed this work by studying a machine that made beams out of composite materials. SOURCE: Courtesy of NASA.

FIGURE 1.13 Artist concept of the Grumman "Beam Builder (B2)" device in space. Although a mockup of the Beam Builder was built and tested in the underwater extra-vehicular activity (EVA) simulator at Marshall Space Flight Center, the work stopped by the early 1980s, and NASA focused on on-orbit assembly of completed parts rather than in-space manufacturing. SOURCE: Courtesy of NASA.

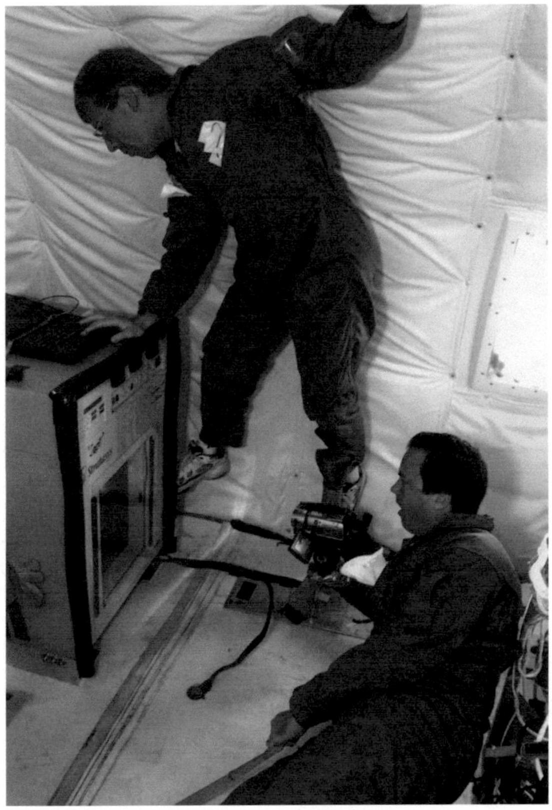

FIGURE 1.14 A 1999 test aboard a KC-135 "Vomit Comet" of an original fused-deposition modeling device. These flight tests were conducted by Cooper and Griffin. This early work, sponsored by NASA's Marshall Space Flight Center, was the forerunner to the upcoming International Space Station experiments. SOURCE: Courtesy of NASA.

Cooper and Griffin foresaw, discussed, evaluated, and anticipated the value of additive manufacturing in microgravity in space enabled by CAD tools and FDM, available at that time via a plastic build material and a 3D printer developed by Stratasys, Inc. Cooper and Griffin wrote, "This task [a report of their work] demonstrated the benefits of the FDM technique to quickly and inexpensively produce replacement components or repair broken hardware in a Space Shuttle or International Space Station (ISS) environment."[27]

With NASA MSFC funding, Cooper and Griffin conducted laboratory experiments and many low-gravity KC-135 low-gravity aircraft experiments. These demonstrated the capability of then-existing FDM equipment to fabricate small test articles in a microgravity environment. Their original device is shown in Figure 1.14. Cooper and Griffin then developed a hardware implementation plan for using FDM for further experiments aboard the ISS, whose initial operations were still in the future. Based on the results of their experiments, they proposed using an ISS FDM device with a 10 cm × 10 cm × 10 cm working volume, a total instrument mass of approximately 45-65 kg with a physical envelope of 0.45 m × 0.5 m × 0.6 m using a peak power of 300 W with air cooling of 150 W.[28] With respect to operations, they stated, "Direct ground control for the input of configuration files and real-time adjustments to the hardware operational parameters would be advantageous to maximizing science return and reducing crew loading."[29]

[27] Cooper and Griffin, *Microgravity Manufacturing Via Fused Depposition*, 2003.

[28] Ibid.

[29] Ibid.

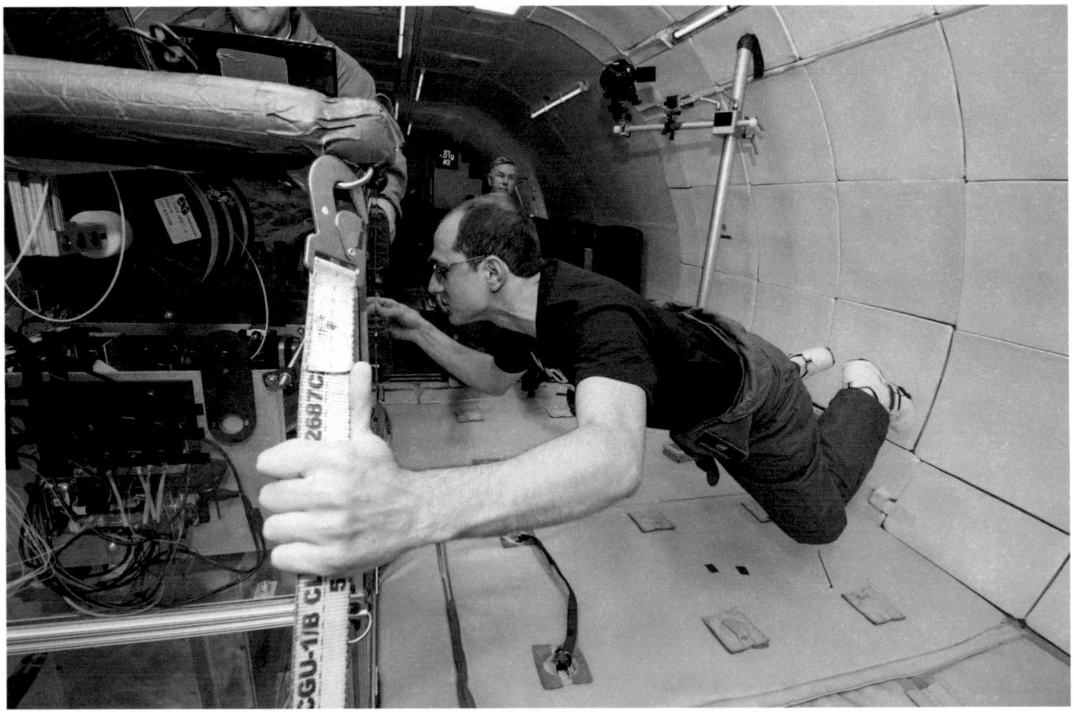

FIGURE 1.15 3D printer developed by Made In Space, Inc., for testing aboard the International Space Station undergoing microgravity testing aboard a parabolic aircraft in 2013. SOURCE: Courtesy of Made In Space, Inc.

In 2010, the start-up company Made In Space, Inc., of Moffett Field, California, was founded and secured funds to build, test and continue the microgravity flight-testing of custom-built and commercially available extrusion additive manufacturing machines with the assistance of NASA's Flight Opportunities Program. Two aircraft flight campaigns in 2011 and 2013 verified and extended the earlier work by Cooper and Griffin which led the company to develop their own space-qualified 3D printer with a NASA technology readiness level of 6, shown in Figure 1.15. This device is scheduled for testing aboard the ISS in late 2014.

## CONCLUSION

Additive manufacturing holds the potential to extend manufacturing capabilities to physical scales currently unobtainable with current spaceflight hardware construction practices. Manufacturing in space may make possible the construction of structures—and possibly entire subsystems—that are fully optimized to the zero-gravity environment, yielding volume-to-mass efficiencies that go well beyond what is attainable at present. The impact of this technology may revolutionize approaches to design.

# 2

# The Possibilities

As part of the study, the committee interviewed government, industry, and academic subject-matter experts on the topic of additive manufacturing. This chapter captures opportunities for space-based additive manufacturing currently envisioned by various groups. The committee offers no opinion on these ideas and neither endorses nor validates the proposed concepts and approaches.

There has lately been great public and industry interest in additive manufacturing because of some progress in laboratory experiments, and some ingenious demonstrations. It is discussed in the general press and popular science magazines and is attracting private investment by manufacturing firms interested in cost savings, as well as by many new small and startup businesses interested in additive manufacturing's compelling possibilities. Academia and government laboratories are also planning experiments. But the technology is still in its infancy and could greatly benefit from well thought out technology roadmaps, standards of quality and performance, common terminology, and other professional engineering standards.

With this as background, this chapter serves as a panorama of possibilities that NASA and the Air Force are considering, but not an endorsement of their likelihood or feasibility.

## CREATING REPLACEMENT COMPONENTS IN SPACE

The most immediate application of additive manufacturing in space relates to creation of replacement parts and components. Data show that a significant percentage of hardware failures on the International Space Station (ISS) involve plastics and composites that may be suitable for repair using additive manufacturing techniques (Figures 2.1 and 2.2 and Box 2.1). This is an area where additive manufacturing could play a significant role. Instead of carrying additional, redundant components to the ISS, parts can be manufactured as needed.

NASA has several efforts under way to bring this vision to fruition.

- A contract with Made In Space, Inc., to verify extrusion-based additive manufacturing in microgravity. The contract's objectives are to print 21 parts (e.g., ASTM standard test coupons, ISS tools, and other parts) that will later be studied on the ground. To date, engineering test units of the printer have been delivered to NASA and have passed environmental tests, and the flight unit is in production. Although this is an experimental platform, the company plans to develop an operational version for operation on the ISS starting in 2015.

- An optical scanner to verify the integrity of parts.
- Research on trade/systems study for metals-based additive manufacturing.

In addition to U.S. efforts, the European Space Agency and member states have similar plans. In 2017, the Italian Space Agency plans to take an fused-deposition-modeling printer on the ISS.

## RECYCLING IN SPACE

In addition to fabricating components, additive manufacturing may present new opportunities for recycling (Figure 2.3). Current NASA efforts include a 2014 Phase I Small Business Independent Research (SBIR) call entitled "Recycling/Reclamation of 3-D Printer Plastic for Reuse." NASA is currently considering eventually transitioning from SBIR to ISS Technology Demonstration in conjunction with planned additive manufacturing activities.

Recycling and reusing materials on the ISS might have a significant impact on space station operations. Currently, astronauts pack trash into robotic spacecraft such as the Russian Progress and the Orbital Sciences' Cygnus for disposal. The spacecraft are detached from the station and burn up in the atmosphere. But before that happens, astronauts spend a considerable amount of time moving the trash, simply to get it out of the way. Using recycled materials in additive manufacturing might ease this logistics and operations problem. Both the component creation and recycling scenarios offer the ability to launch feedstock in bulk instead of delicate hardware. However, separating products and materials is a large portion of any recycling effort. On a vehicle like the ISS, this may be a time-consuming manual task. Clearly this will require careful evaluation to determine if it is the most efficient solution.

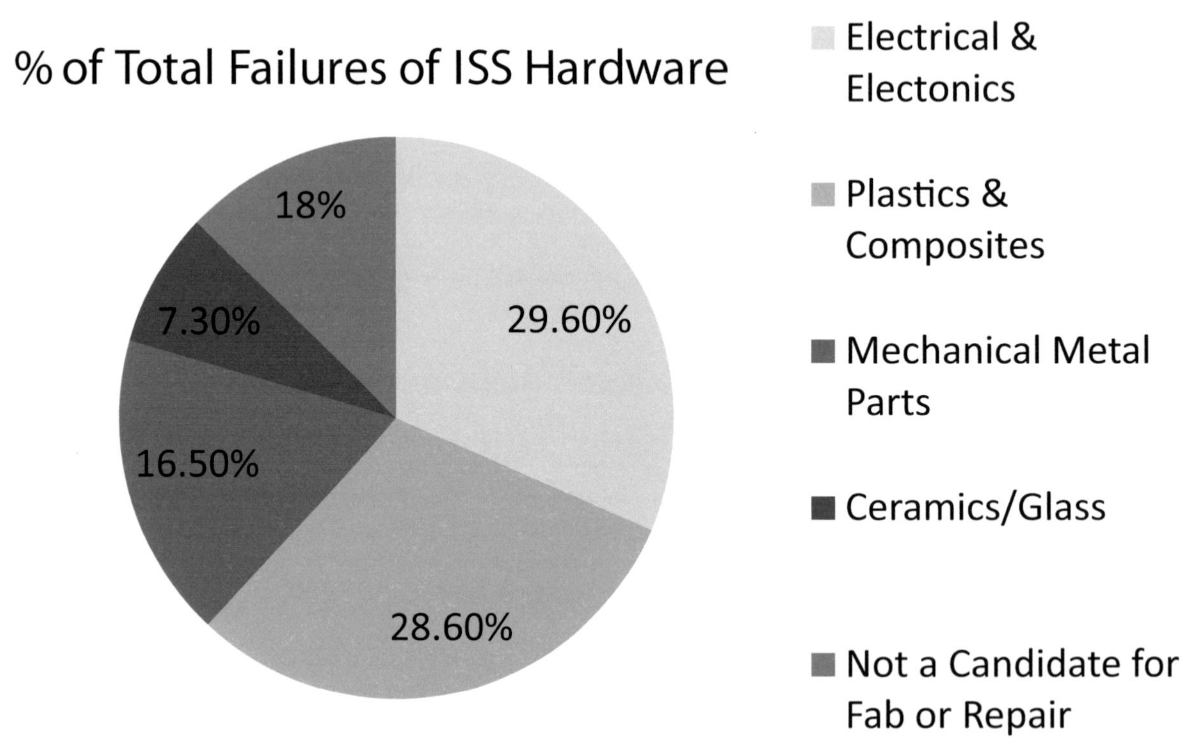


FIGURE 2.1 The percentage of failed parts and components on the International Space Station (ISS) that are candidates for repair or fabrication divided into categories. Some of these could in the near term conceivably be replaced with additively manufactured parts made on the ISS. SOURCE: Courtesy of Made In Space, Inc.

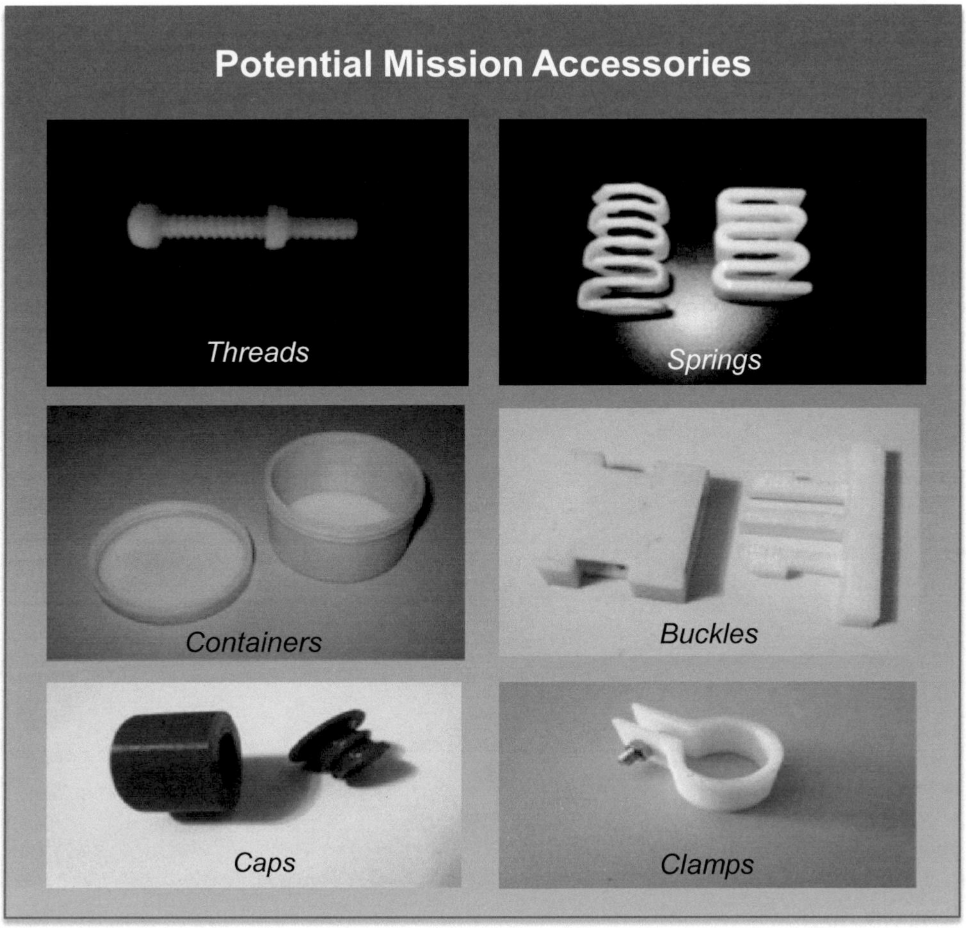


FIGURE 2.2 Parts that can be additively manufactured. SOURCE: Courtesy of NASA.

## REPLACEMENT COMPONENTS FOR ROBOTIC SPACECRAFT

In contrast to the possibility of creating replacement components onboard human spacecraft, the prospects for producing replacement components for robotic spacecraft in orbit are far less clear. For robotic spacecraft, up to 50 percent of the failures are attributable to power subsystem failures.[1] In 1992, a survey was published of 2,500 spacecraft failures that took place between 1962 and 1988. About 50 percent were identified and traceable to issues encountered in operations, the space environment, or with design problems. About 30 percent of the failures were random, likely due to manufacturing and workmanship. In nearly 19 percent of the failure cases, causes could not be determined, perhaps indicating that system telemetry was inadequate.[2] Given the complexity and types of failures, it is difficult to envision how in-space additive manufacturing would be able to successfully contribute to the manufacture of replacement parts for robotic spacecraft.

[1] "Commercial Communications Satellite Bus Reliability Analysis," Frost & Sullivan, August 2004; D.M. Harland and R.D. Lorenz, *Space System Failures*, Springer-Praxis Publishing, Chichester, U.K., 2005.

[2] H. Hecht, Reliability during space mission concept exploration, in *Space Mission Analysis and Design* (W.J. Larson and J.R. Wertz, eds.), Microcosm, Torrance, Calif., 1992.

**BOX 2.1**
**Emergency Repairs in Space**

FIGURE 2.1.1 The adapter assembled by the Apollo 13 astronauts from equipment inside their spacecraft including pieces of folders, plastic bags, and duct tape. Such a device could be produced by an additive manufacturing machine today. SOURCE: Courtesy of NASA.

In 1970, the Apollo 13 Command Module suffered a catastrophic failure on its way to the Moon. The crew encountered many risks to their survival during the mission. They were forced to use the Apollo Lunar Module as a "lifeboat" to keep them alive during the return trip. But the lithium hydroxide canisters for the Command Module were different from the ones for the Lunar Module and did not fit in the same receptacle. With the assistance of engineers on the ground, the astronauts developed a makeshift adapter to connect the two. In 2013, an engineer from the firm Made In Space, Inc., spent 1 hour designing an adapter for the lithium hydroxide canister and was able to print it and demonstrate its operation by the end of the day (Figure 2.1.1). This experience demonstrates the flexibility and adaptability of this technology to address unforeseen situations.

## CREATE STRUCTURES DIFFICULT TO PRODUCE ON OR TRANSPORT FROM EARTH

Additive manufacturing is not simply a way to produce the same parts in a different way. Much more importantly, it enables users to design parts for ultimate use rather than for their machining qualities. In other words, additively manufacturing hardware in space could enable production of ultra-low-mass systems and parts for use, thereby easing stowing and launch requirements. Currently, large components and systems such as antennas, booms,

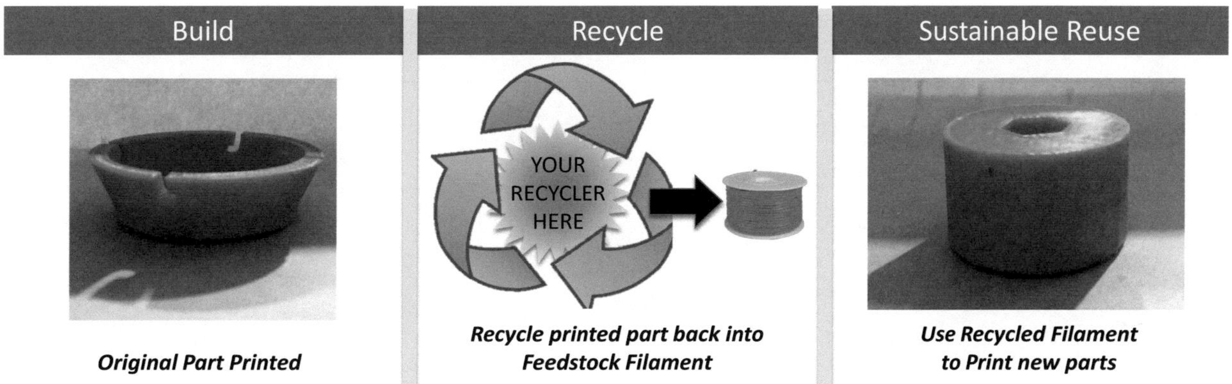


FIGURE 2.3 Parts that can be recycled for new additive manufacturing. SOURCE: Courtesy of NASA.

and panels are designed for launch. Their delicate structures, sizes and shapes are limited by the requirement to stow them within available launch fairings. This severely limits functionality with respect to scaling. Accommodating these structures involves trade-offs between launch vehicle lift capability and shroud size. This problem is often solved by using a larger-diameter shroud that imposes a mass penalty.

While deployable structures have enabled construction of large systems, their packing efficiency is not sufficient to enable the kind of kilometer-size scaling required for many applications such as long-baseline interferometry and sparse aperture sensing.[3]

With the use of additive manufacturing and supporting technologies in space, on-orbit construction and "erectables" technologies can enable deployment of systems that need not conform to weight and volumetric constraints posed by launch fairings and shrouds.[4] Structures envisioned include ultra-thin mirrors, gossamer structures like ribbons, large antennas and arrays, reflectors, and trusses, among others. Such structures can be envisioned to better enable exploration (e.g., provide better sources of power for long-duration exploration activities).

The vision for building such structures involves launching raw materials in a compact, durable state together with software for fabricating, assembling, and integrating components to an already on-orbit additive manufacturing machine that will manufacture the new operational space system. NASA is currently funding Tethers Unlimited, which proposes to take compact materials that are easy to launch, such as spools of thread, to form large truss-based structures such as kilometer-long solar arrays and antennas in space. The company claims that by constructing these structures on-orbit, they can be made with much lower tensile strength requirements; nor do the structures need to survive the harsh vibrations of launch and deployment. The firm has identified several applications, including a star shade to block light from stars so that a space-based telescope can image exoplanets around those stars. This would involve the deployment of a large "SpiderFab," a spider-like robot that can "extrude" long beams and join together large structures (Figure 2.4).

## CREATE SENSORS, SENSOR SYSTEMS, AND SATELLITES

Additive manufacturing in space could potentially enable production of not just components, but also entire subsystems and systems. An example the committee was briefed on included production, assembly, and launch of sensor-loaded CubeSats from the ISS or other platforms in orbit.

---

3 Adapted from Robert Hoyt's NASA NIAC Phase 1 report, available at http://www.nasa.gov/directorates/spacetech/niac/2012_phase_I_fellows_hoyt_spiderfab.html#.UvZAofldWtY.

4 Such structures have been built today without additive manufacturing. For example, the ISS was built on-orbit by assembling multiple components launched separately. Unfortunately, the cost of multiple launches and the labor required for such assembly is high enough that the ISS is not an ideal model for deployment of large space-based structures. ISS assembly required, for example, 89 Russian and 37 shuttle launches, 168 spacewalks spanning 1,061 hours, transport of 924,739 lbs of material, and 2.3 million lines of computer code.

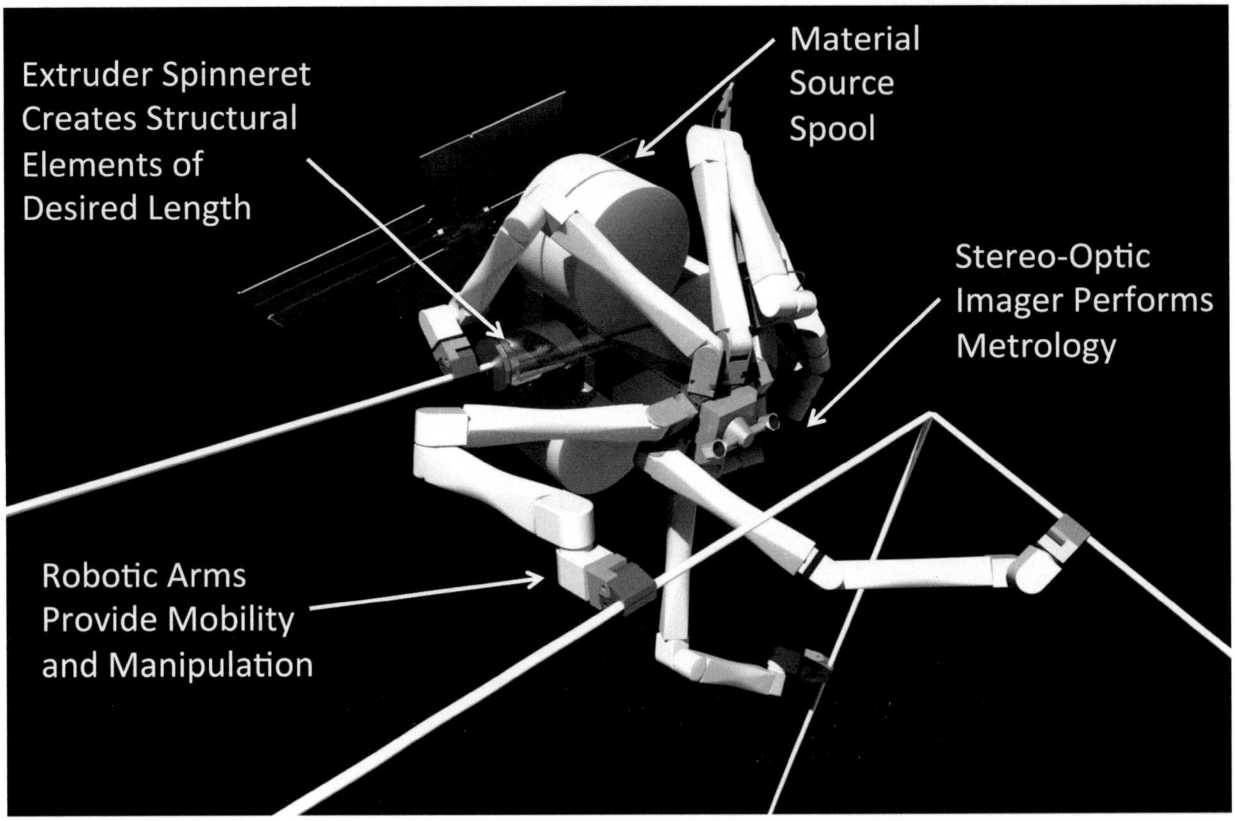


FIGURE 2.4 SpiderFab, a combination of a machine that creates structural elements and a multi-dextrous robot that can manipulate those elements. SOURCE: Courtesy of Tethers Unlimited.

Additively manufactured in-space satellites deployed not only singly, but also as "swarms" have also been envisioned. The proposed Automated Manufacturing Facility on-board the ISS or a standalone CubeSat platform could build swarms of satellites. The swarm could have a range of capabilities, and possibly even act as a fully functional satellite system.

Additive manufacturing research and development is under way in several Air Force Research Laboratory (AFRL) directorates as well other parts of the Department of Defense. The Defense Advanced Research Projects Agency (DARPA) is also exploring additive manufacturing in space. For example, the DARPA Phoenix program is developing and demonstrating technologies to harvest and reuse valuable components from retired, nonworking satellites in geosynchronous orbit and to demonstrate the ability to create new space systems at greatly reduced cost (Figure 2.5). The program envisions developing a new class of small "satlets" that could be sent to the geosynchronous orbit region as a "ride along" on a commercial satellite launch. The "satlets" would then attach to the antenna of a non-functional cooperating satellite robotically, essentially creating a new space system.

## FREE-FLYING "FAB LAB"

The concept of a fabrication laboratory or "fab lab" was developed at the Center for Bits and Atoms at the Media Laboratory at the Massachusetts Institute of Technology to explore how the content of information relates to its physical representation. A typical fab lab is equipped with an array of flexible computer-controlled tools, often including additive or 3D printers, with the aim to make "almost anything," in the words of its creators. The concept

FIGURE 2.5 Artist illustration of the Defense Advanced Research Projects Agency's (DARPA's) Phoenix program. In this illustration a robotic spacecraft is attaching new components to an antenna harvested from a decommissioned spacecraft. SOURCE: Courtesy of DARPA, see http://www.milsatmagazine.com/story.php?number=562432496.

of a fab lab in space is similar: users—human or robotic—would be able to access tools in space to manufacture what is needed without bringing it from Earth. An additive manufacturing capability would be the heart of such a fab lab. There are several free-flying spacecraft that are either currently available or will probably become available within the next decade that could serve as free-flying fab labs for additive manufacturing in space.

## FULLY PRINTED SPACECRAFT

Additive manufacturing technology can be applied to subsystems as well as entire spacecraft and can be useful even when what it produces does not meet the conventional definition of a spacecraft. An example of a two-dimensional sensor being developed at the NASA Jet Propulsion Laboratory, funded by the NASA Innovative Advanced Concepts (NIAC) program, is shown in Figure 2.6. This is not a "spacecraft" by common definitions. It is essentially a transparent sheet of plastic with printed electronics that has been proposed to collect environmental data in space or in a planet's atmosphere. It demonstrates that technology can change conventional understanding of what is possible.

Some people have suggested that additive manufacturing in space could—autonomously or with human

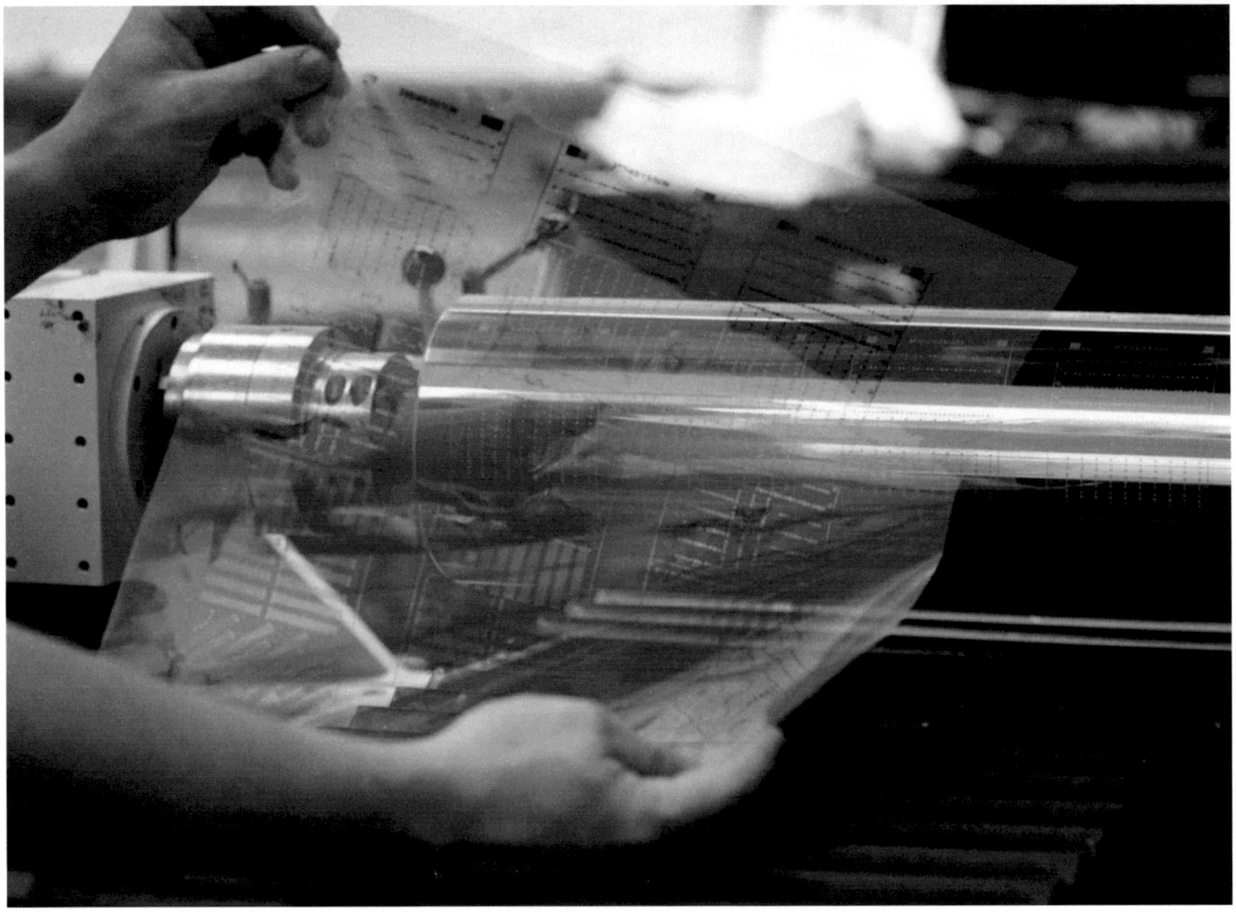

FIGURE 2.6 A two-dimensional printed spacecraft being developed at the Jet Propulsion Laboratory. SOURCE: Courtesy of PARC, a Xerox company; http://gigaom.com/2013/08/20/nasa-wants-to-print-a-spacecraft-but-first-its-printing-the-electronics/.

support—create not just components but an entire spacecraft in space. The spacecraft could be built as a single unit with a single machine or assembled by humans or even autonomously. How long it would take to realize this vision depends not only on technical advancements made but also on how a spacecraft is defined. A single-function spacecraft (one that, for example, only measures solar radiation during a space weather event and then degrades) is feasible on a shorter timeline than a multiple-function spacecraft that is radiation and nuclear hardened, intended to last multiple years, made of multiple materials, and which serves many functions. The latter is likely many decades away.

## USE OF RESOURCES ON PLANETARY SURFACES

Availability of construction material (e.g., metals, water) in space (e.g., on asteroids or on surfaces of planetary bodies) enables the possibility of additively building settlements and other facilities without having to take expensive and bulky prefabricated materials out of Earth's gravitational field. Lunar regolith, for example, could be used to construct pressurized habitats for human shelter as well as other infrastructure (e.g., landing pads,

FIGURE 2.7 A robot on the Moon using "contour crafting" to build up a structure, layer by layer. SOURCE: Courtesy of Behrokh Khoshnevis, University of Southern California.

roads, blast walls, shade walls, and hangars for protection against thermal radiation and micrometeorites) on the Moon (Figure 2.7).

Another NIAC grant to a team at the University of Southern California is exploring the concept of using a technique called contour crafting to build infrastructure (landing pads) on the Moon using simulated lunar regolith. This technique extrudes a material such as concrete layer by layer to build up structures such as walls. The European Space Agency is funding similar research with a technology called D-Shapes to design a Moon-based habitat (Figure 2.8).

## SUMMARY

Additive manufacturing might provide the means to transform space system architectures. Space system configurations that are now dominated by requirements to survive ground manufacturing, assembly, test, transport, and launch could be reexamined as this new capability becomes available. Relaxation of volume limitations of a launch vehicle shroud, which currently place restrictions on the physical size of a spacecraft, could enable structures beyond what is presently attainable. Structural designers of spacecraft could have an entirely new set of implementation approaches for spacecraft configuration that might not need to account for loads and accelerations in the launch vehicle ascent process.

A launch vehicle transporting material for additive manufacturing in bulk could deliver a payload to space with volumetric densities up to 100 times that currently attainable. Launching materials in bulk would enable more economical ballistic departures.

A space-based manufacturing center tailored to the production of spacecraft on orbit will need to be conceived that will enable designers of tomorrow's spacecraft to create a digital concept that the space-based manufacturing center would transform into a functioning machine, using the bulk material, tailored to the zero-gravity environ-

FIGURE 2.8 Additively built lunar settlement. SOURCE: Courtesy of the European Space Agency and Foster + Partners.

ment. Full software simulations in the hands of a new generation of scientists and engineers could enable digital simulation of novel configurations for systems to be constructed in space.

The process of developing spacecraft in support of space missions—a process that currently extends up to a decade in length for large spacecraft—might be transformed through the use of (as yet) undeveloped design and construction tools tailored to the additive manufacturing process. Additive manufacturing holds the promise of relaxing current constraints on physical shape and size—opening up the opportunities for physical scales (both large and small) that may be approached by designers in entirely new ways.

The following chapters will explain that the possibilities of additive manufacturing in the near-term are modest—creating replacement components, recycling parts into feedstock, etc. However, in the long run, if near-term efforts are carefully designed and executed, the knowledge base to functionally reconceptualize space architectures could be developed. The application of additive manufacturing to the space environment could lead to a change in ideas and concepts of what satellites look like, how they are designed, and what functionality they have.

The rest of this report examines the possibilities discussed above in light of current technology capabilities and trajectories and proposes a roadmap of how to get from where additive manufacturing in space is today to where it could be in the next 20-40 years.

# 3

# Technical Challenges for the Use of Additive Manufacturing in Space

Additive manufacturing is a technology that is being enthusiastically pursued across many fronts. While there are many technically feasible approaches to additive manufacturing of parts and systems, not all are adaptable to space. Chapter 2 highlighted a vision, along with some specific scenarios, of what additive manufacturing in space may enable. This chapter will outline difficulties the committee sees with reaching the identified scenarios.

Additive manufacturing is widely used but not yet heavily used for production. There is a lot of research and development work aimed at trying to perfect and understand the various additive manufacturing techniques, expand the type of materials and parts that are utilized, and explore the manufacture of a complex, integrated system. However, there are still many questions that need to be addressed before additive manufacturing can be utilized in a production mode for critical aerospace applications. Before moving additive manufacturing technology to the space environment, further development in several fundamental areas needs to be complete and well understood. These areas represent barriers to wider use, even in a ground-based environment. (Figure 3.1 depicts a recent demonstration of the additive manufacturing of a rocket engine thrust chamber.)

## MATERIALS DEVELOPMENT AND CHARACTERIZATION

Although a wide range of homogenous and heterogeneous material mixtures are employed in additive manufacturing, there is still a need for developing targeted materials for use with each specific technique. New physics-based models of additive manufacturing processes are needed to understand and predict material properties and help optimize material composition. A better understanding of the basic physics could then potentially lead to predictive modeling, allowing designers, engineers, scientists, and users to estimate the functional properties of the part during design and tweak the design to achieve desired outcomes.

## PROCESS MODELING AND CONTROL

Methods are needed for in-process monitoring and closed-loop feedback to help improve consistency, repeatability, and uniformity across machines. In situ sensors are an area that holds potential for providing nondestructive evaluation and enabling early defect detection, particularly related to thermal control. Of particular importance is gaining a better understanding of the processing, structure, and property relationships to fully understand the characteristics of the final product. Given the same starting material, different processing approaches can affect

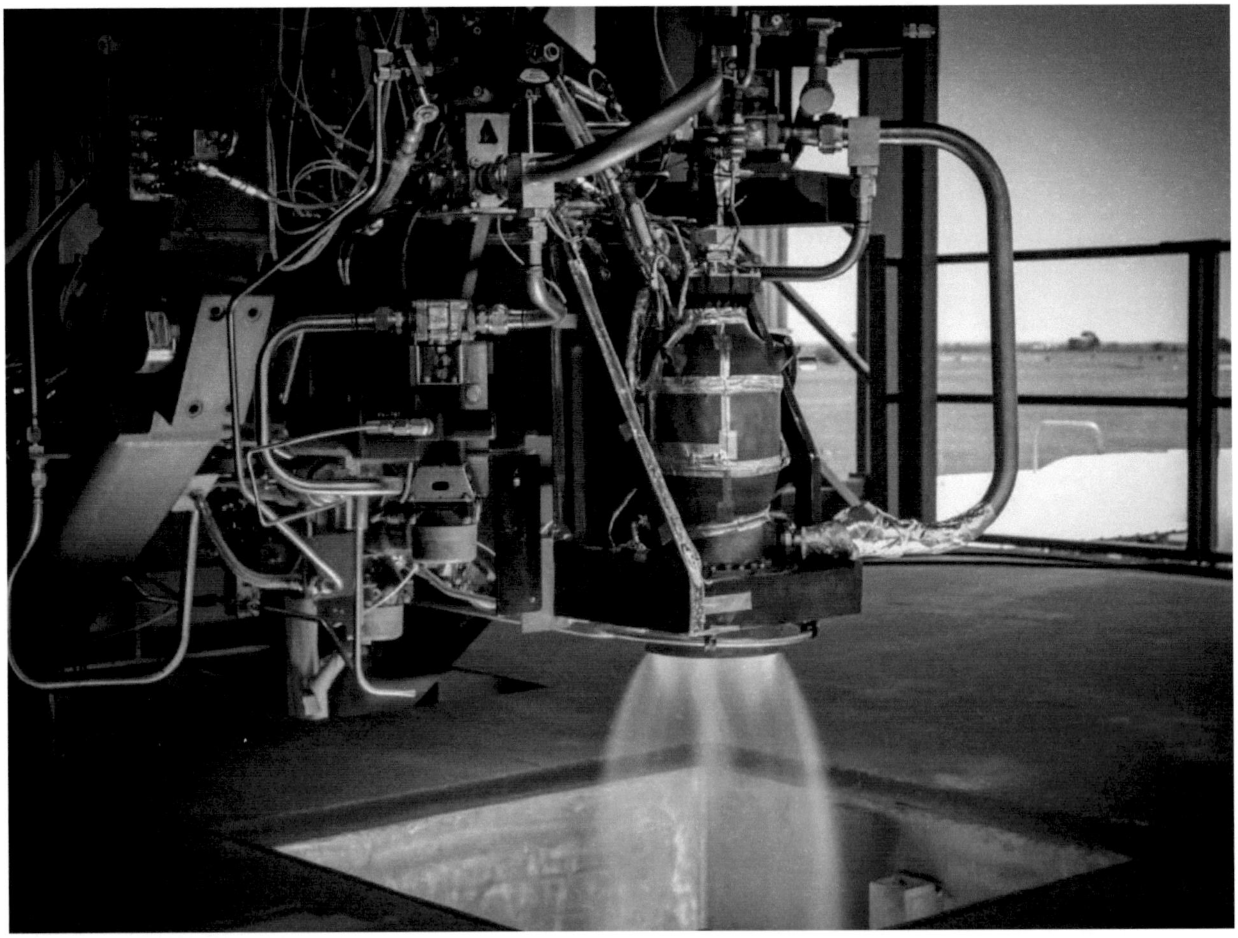

FIGURE 3.1 Test of a SuperDraco rocket engine at full power using an additively manufactured Inconel thrust chamber. SOURCE: Courtesy of SpaceX.

the properties of the final product in different ways. For example, the thermal effects can vary depending on the energy source, energy density, and environment in which the process occurs.

Although some of this report focuses on physical prototyping by additive manufacturing, the role of virtual prototyping by process modeling, simulation, and open-source data simulations is also important. A deep level of 3D planning, simulation, and cross-platform analysis can optimize the to-be-built system prior to the expensive creation of parts by additive manufacturing. The opportunity for NASA and the Air Force is to invest in systems that produce open-system design, planning, simulation, and analysis tools. Some reports indicate that substantial savings can be obtained using a design-flow process: a 30 percent reduction both in overall production cost and time to market; 25 percent savings in plant and facility layout; 30 percent cost savings in labor utilization; 35 percent cost savings in optimized material flow; and 15 percent savings in improved quality from validation of processes prior to production.[1] Recently, Boeing designed the 787 aircraft through extensive use of such simulation software. Multi-physics (e.g., computational fluid dynamics and structural) codes were used to replace extensive wind-tunnel testing of proposed physical models. Once the codes produced an optimized design, a physical model was built and tested. The resulting design is fuel efficient (fuel savings of up to 30 percent are expected) because

[1] C. Collier, Improving wind blade manufacturability, *Energy Manufacturing,* 2011, pp. 61-63.

of the lighter-weight carbon-fiber construction. Boeing also claims that this has resulted in reduced design time, which has undoubtedly resulted in energy and cost savings resulting from fewer wind-tunnel and other physical testing. Additive manufacturing is only one step in the design-flow process (focused on the rapid prototyping step). Significant savings and optimizations should be done prior to additive manufacturing.

## PRECISION AND RESOLUTION

While different additive manufacturing technologies can produce different levels of precision and accuracy in the final part, all techniques have limits as compared to more traditional subtractive techniques. Resolution describes the smallest features that the device can print. Resolution is often defined in units of reciprocal length, for example, 20 spots per centimeter. It is useful to think of resolution in unit-less terms, citing the total number of resolved positions over the full range of an axis. A resolution of 20 spots per centimeter over 25 centimeters resolves 500 distinct positions for the print head. This gives a better idea of how clearly a printer can define an object. For contrast, a television image resolves about 720 pixels horizontally, while a xerographic printer can resolve 24,000 spots over a 10-inch page, producing a correspondingly better image.

Accuracy describes how precisely the printer can deposit material in separated places. If the printer is asked to make a cube exactly 10 centimeters on a side, how big will the result actually be? Will its surfaces be perpendicular to each other? Will the diagonals of the cube be equal in length? It is easy to assume that the printer's accuracy is equal to its resolution, but material shrinkage, axis misalignment, and servomechanism error may affect accuracy. For example, no one expects the image printed by a xerographic printer to faithfully retain the exact size of the original. Paper shrinkage that occurs in xerographic printing limits accuracy but not resolution.

In additive manufacturing, the direction of the beam or the processing direction can induce thermal stresses in the part, impacting the precision of the final part. The resolution of the beam and/or particles also influence heat transfer and internal stresses, thereby influencing the integrity of the final part.

## CONSTRUCTION TIME CONSTRAINTS

In general, three-dimensional printers create parts in several ways, using solid sheets or fibers or by bonding material using a laser. When the material is added one spot at a time, the time required to print an entire object depends on how many spots the printer must print. The total volume required to print therefore can affect the time of the build process. For example, hollow objects or web-like objects print faster than solid objects because they require fewer spots. The time to print an object also depends on the resolution. Printers with finer resolution have to process more spots for a given output volume than printers with coarse resolution. The trade-off is that the finer resolution produces a smoother rendition of the product at the cost of additional time. Hence, the complexity of the design, and perhaps application of the part, will also play a role in the time required to manufacture it. A requirement for multi-material manufacturing capability will also impact time of construction.

At the current state of the art, CubeSats and similar small satellites can potentially be constructed over several weeks if printing is required for various complex subsystems. However, even simple, monolithic metal objects of masses greater than about 1,200 kg built with additive manufacturing techniques with the finest resolution require a full year to fabricate. Based on data from current additive manufacturing manufacturers, use of other materials such as plastics or composites will require even more time than metals at these resolutions. Design for and construction of an object in space will likely require much less mass, due to the reduced gravity, but it is difficult to predict the corresponding impact on time of construction without knowing the resolution required and the impact of other environmental effects on the process. Laying down a lot of material for metals or plastics can be done in a short period of time at relatively coarse resolutions; however, the higher resolutions required for satellite components would result in longer fabrication times.

## DESIGN TOOLS AND SOFTWARE

Solid-modeling software and computer-aided design (CAD) are required to use additive manufacturing technologies. Along with appropriate design software comes the need for standardized file formats for communication between the design and manufacturing environment. The STL (stereolithography) format is now the standard for communication across additive manufacturing machines. However, new CAD tools are needed that can simultaneously optimize both shape and material properties and design complex lattice structures that optimize reductions in material and weight.

## MACHINE QUALIFICATION, CERTIFICATION, AND STANDARDIZATION

Machine qualification standards could help machine-to-machine and part-to-part repeatability. Along with a standardized material properties database, qualification at a machine or process level could help assist with the final product qualification and verification requirements.

As important as the contributions and prospective technical improvements associated with additive manufacturing may be, the technology will not be comprehensively deployed for production purposes without intensive study and approval of definitive standards. Space hardware must undergo stringent certification processes due to the complexity of operations, the strenuous environmental conditions that hardware must operate in, and safety considerations. Thus, space system engineering is a very mature, complex discipline with conservative approaches to all programmatic and technical changes.

Certification is of critical importance to the progress of additive manufacturing for use in the aerospace industry. The aerospace industry cannot yet leverage the potential benefits of additive manufacturing until stringent qualification and verification requirements are met. Robust, well-understood certification approaches will need to be developed for additive manufacturing technologies. Currently, material properties and structural design are not yet uniform or standardized. Testing the final products, which is the current method of quality assurance and verification, requires additional time and resources. A more efficient, systematic part-certification process is needed. Closed-loop process-control systems can quantify inconsistencies allow for real-time quality control. New sensors used in conjunction with closed-loop control systems can document in situ data on precision, surface finish, porosity, melt pool size, and other parameters. Methods for inspecting the build environment during processing may also be required in order to make corrections as needed.

**Finding:** There are some fundamental issues that need to be resolved concerning additive manufacturing and its utilization for terrestrial purposes before a space-based application can be derived.

- A clear understanding of structure-property relationships and their dependence on processing techniques needs to be established to ensure consistency in production. The production process can also benefit from standardization of design software, file formats, and processing and equipment parameters. Most importantly, a verification and certification methodology will have to be defined that guarantees the quality of the additively manufactured parts.
- Aerospace systems have critical missions and have to meet rigorous standards for quality and reliability, with standards that are set to ensure mission success.
- In order to benefit, even terrestrially, from additive manufacturing approaches, the issues of qualification and certification will have to be addressed. A standard approach to qualification and certification of finished parts will simplify the application of additive manufacturing to the space environment and also enable more widespread application on Earth.

## ADDITIVE MANUFACTURING AN ENTIRE SPACECRAFT ON THE GROUND

Current approaches to complex, multi-material, and multi-functional additively manufactured parts can involve embedding a circuit board, motor, or other subassembly into the process when and where it needs to be integrated.

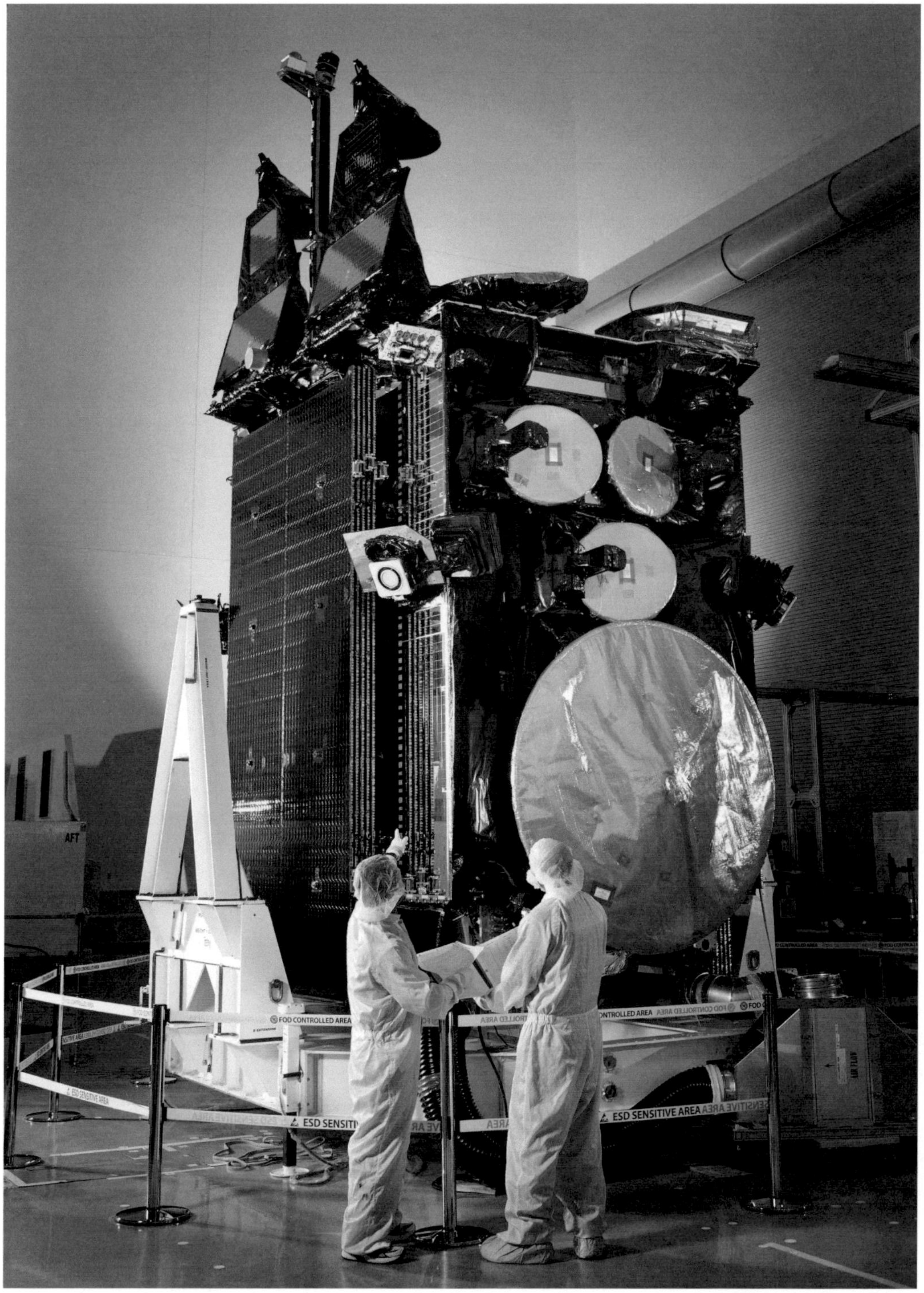


FIGURE 3.2 Lockheed Martin Advanced Extremely High Frequency communications satellite built for the Air Force. Many satellites constructed for operational military and civilian missions are large and complex. Although they may incorporate additively manufactured components, the next major challenge will be additively manufacturing major subsystems on the ground. SOURCE: Courtesy of Lockheed Martin Corporation.

(See Figure 3.2.) This can be done easily for ground-based, human-tended systems: The process is halted, the sub-part is introduced to the environment, and the manufacturing process restarted. Complex final parts are thus achieved in a cost-effective and efficient manner. To manufacture a complete functional satellite on orbit in an environment that may not have the same level of human tending, it is likely that a strong "pull" from the government would be required to encourage research in additive manufacturing in these areas.

Efforts are currently underway on the ground to apply additive manufacturing to the production of a spacecraft. The briefings to the committee on the state of the art revealed two categories of effort currently under way, including (1) additively manufactured spacecraft structure and (2) additively manufactured structure with embedded electrical conductors and components. The committee did not see any evidence that companies are considering a complete, monolithic, additively manufactured spacecraft on the ground. This would be an objective that requires significant development work.

The most advanced ground-based work is focused on the manufacture of structural components to reduce costs. Commercial companies such as Lockheed Martin Corporation, Aerojet Rocketdyne, Orbital Sciences Corporation, and others are working to expand the use of additive manufacturing beyond just structure to a broader number of components, and there is promising work in the development of an integrated structure and propulsion system at machine scales (a cube 0.5 m on a side). The work at the component level is an important step to understanding the application of additive manufacturing to full spacecraft manufacturing.

Lockheed Martin, Northrop Grumman, Stratasys, RP+M, University of Texas, El Paso (UTEP), University New Mexico, and Youngstown State University together funded by America Makes, are undertaking laboratory work to additively manufacture structures with embedded electric conductors, providing a way to integrate and connect electromechanical subsystems into the assembly. The integration of a wiring harness within the structure represents a significant technological step beyond structure manufacture, and when this technology is fully developed, it will be one of the keys to the complete manufacture of a spacecraft and its subsystems using additive manufacturing. Government research and development investment spanning decades has been required to advance this capability. Building wiring harnesses and distribution networks integral to space structures has been an ongoing Air Force Research Laboratory (AFRL)/Space Vehicles Directorate endeavor since the 1990s, with the most recent program being the PnPSat platform. The challenge of combining multiple elements into a monolithic build has typically resulted in higher costs, more complexity, and more risk. Coupling two subsystems or functions together results in interdependencies where a single manufacturing defect or anomaly in one function results in the overall failure of the combined system (during satellite assembly). For example, an acceptance testing failure of an embedded wire would result in scrapping the entire part, not just replacing a wire, even though the strength and stiffness of the structure may be unaffected.

Advancing to the state of producing a complete spacecraft on the ground through the additive manufacturing process will require specific developments in each of the spacecraft subsystem areas. The following sections describe some of the technical challenges in advancing additive manufacturing with the ultimate goal of manufacturing a complete spacecraft. Significant new technologies will be needed to do so, particularly in electronics and optical manufacture where mirrors and optics are fabricated using photolithographic and surface figure techniques at scales that are orders of magnitude better than the most refined of additive manufacturing processes.

Work is currently under way in four subsystem areas: structure, thermal, propulsion, and power. Spacecraft structures are being constructed, and most importantly, a repository of knowledge of how to create structural components using additive manufacturing is currently being developed. Commercial interests to accelerate production, while realizing cost economies, are driving advances in this area. Characterization of the thermal performance of materials (thermal conductivity) as well as optical properties (solar absorptivity and infrared emissivity) is being carried out as part of the structural design development process. Additively manufacturing an external spacecraft thermal control system, using radiatively coupled surfaces, will benefit from the further development of metallic and nonmetallic materials. The development of propulsion system components that include nozzle parts, as well as tank and fuel feed systems manufactured as one component, are currently being produced at CubeSat and larger scales. The development of electromechanical valves will be required to fully manufacture a propulsion system (absent propellant). Spacecraft power system work is proceeding in printing photon-to-electron conversion devices (solar cells), power storage devices (batteries), and electrical conductors embedded in the additive manufacturing process. Based on briefings presented to the committee, the development of these four subsystems is proceeding, and as separate subsystems, the use of additive manufacturing is being explored. Combining the manufacture of these individual systems into an integrated unit will be one of the next significant steps.

The remaining three subsystems—attitude determination and control, command and data handling, and telecommunications—hold significant technical challenges if they are to be produced through the additive manufacturing process. The control of spacecraft attitude at modest levels of performance requires the use of optical, photosensing, and magnetic sensing systems. Materials development for these purposes is largely unexplored in the additive manufacturing processes. In optics fabrication, surface figures measured in parts of wavelengths of light would have to be developed. Developing additive manufacturing technology for sensors to replicate current capabilities to view the Sun, stars, and infrared signals will be important in this subsystem area. Momentum management devices that include either magnetic coupling or rotating mass will require the development of motors (currently employing rare earth materials) and rotational support structures (currently ball bearings) of sufficiently high precision. Manufacture of ball bearings with an ABEC (Annular Bearing Engineering Committee) quality rating of 5-7 requires precision grinding of balls and bearing races to achieve the necessary surface finish. This quality is not achievable with current additive manufacturing processes (Figure 3.3). The ability to attain precision satellite pointing as a function of frequency (jitter) will be directly affected by the developments in these areas.

FIGURE 3.3 Ball bearings are mechanically and physically simple devices that play important roles in many spacecraft. Nevertheless, they cannot currently be additively manufactured because the technology cannot achieve the required precision. SOURCE: By Androstachys (own work) [CC-BY-SA-3.0 (http://creativecommons.org/licenses/by-sa/3.0) or GFDL (http://www.gnu.org/copyleft/fdl.html)], via Wikimedia Commons.

Current spacecraft command and data-handling systems (the spacecraft computer) are an outgrowth of the development of semiconductor technology, coupled with an understanding of the design of electronic components to survive in the radiation environment of space. Physically replicating electronic components (currently produced by photolithography at scales of 35 nanometers) at currently common additive manufacturing resolution (>50 microns) would produce components 1,000 times larger than the physical size of currently available parts. Due to geometry alone, the power consumption and processing speed would be inferior to current technology. Perhaps more importantly, the development of materials in additive manufacturing that replicate those used in semiconductor processing remains largely unexplored. Advancing from the current state of additive manufacturing of primitive electronics (conductors) with embedded prefabricated components to the development of "thinking" materials (i.e., semiconductors) for logic and memory, at scale sizes small enough to fit within a spacecraft, will be a significant development challenge. Although lesser capabilities may be acceptable for some applications (and can be traded-off for other benefits, such as cost), they will likely not be acceptable for critical or high-value applications. This may prove to be one of the most intimidating technology areas. The ultimate solution may be to use additive manufacturing to produce what is reasonable and place or integrate components produced by other means.

On the other hand, a valuable field of research might emerge from attempts to replace such precise mechanical components with mechanisms suitable for additive manufacture. For example, flexures requiring far less precision might replace bearings. Consider also electric motors such as those required to actuate valves or to position antennas. Motors require a combination of materials with vastly different properties. The coils must have both good conductors and good insulators. No additive manufacturing process has yet been able to mix metals with insulating materials. Motors also require magnetic materials. Little is known about the magnetic properties of materials placed by additive manufacturing, although there is some new development work in this area.[2] As an example, the University of Texas, El Paso has demonstrated 3D printing a brushless direct current motor in a single build sequence, but this approach required placing previously fabricated electronic components during the build.

While additive manufacturing seems well suited to the manufacture of telecommunication antennas, other parts of this subsystem will need development and face challenges similar to those in the command and data-handling system.

**Finding:** Considering that the present state of manufacturing focuses on new types of individual components of specialized shapes, composition, and materials, it is clear that the task of additively manufacturing a complete scientific or military satellite of the complexity of current spacecraft is far in the future. This situation is unlikely to change unless major, very-long-term changes are made in this nation's space systems design, engineering methodologies, and infrastructure at all levels.

## TRANSITIONING ADDITIVE MANUFACTURING TECHNOLOGY TO THE SPACE ENVIRONMENT

There are many fundamental questions that need to be answered before additive manufacturing can become widely applicable for routine Earth-based manufacturing. Once ground-based additive manufacturing technology has matured enough to be a viable process for aerospace applications, a logical step forward is the transfer of the technology to the space environment. However, even with a defined, consistent, qualified additive manufacturing process, new issues will have to be addressed when transferring the technology to space. Not only will some fundamental process-related questions need to be readdressed, driven by the space environment, but questions relating to infrastructure, platforms, and the overall manufacturing approach as a whole need to be answered.

A manufacturing capability in space can be placed in a pressurized, climate-controlled environment or located on an unpressurized platform in the vacuum of space. In either event, the microgravity environment will be a factor. Depending on the location of the capability, the impact of vacuum and thermal environments on the additive manufacturing technology will also need to be considered.

[2] Doug Hofmann, "Lightweight and Multi-Functional Materials & Structures," presentation to the committee, November 12, 2013.

### Space Environment: Gravity and Vacuum

Depending on the technique of interest. the placement of an additive manufacturing capability in the vacuum of space may present either barriers or opportunity. The technique using electron beams for the energy source, for example, is designed for operation in a vacuum. Additional research is needed to prove the concept.

The most fundamental technical issue that will have to be dealt with when moving the additive manufacturing capability to space is the effect of zero-gravity or reduced gravity on the manufacturing process and hence the properties of the final product. Each technology will have different challenges in adapting the processes to an environment where gravity is not available as a control variable. In the absence of gravity, surface tension forces become important determiners of system behavior, and processes that rely on the control of fluid or flow conditions will need further research. The reduced- or zero-gravity environment will not only have an effect on the process parameters and techniques, but will also potentially have an effect on the final mechanical and functional integrity of the finished part.

The lack of gravity will also be a factor in the design of handling and support systems for the product. Current additive manufacturing systems employ linear XYZ drives delivering both accurate and precise motion, and generating this motion on Earth in a 1 *g*, thermally controlled environment is well understood. In addition, designs for drive, bearing, and lubrication systems that yield consistent geometry and predictable feature size are also well understood in 1 *g*. Translating and understanding the required system performance into reduced gravity as well as in a vacuum is critical to controlling the manufacturing process. In addition, in zero gravity, floating debris can damage the product, and the machine, and will have to be controlled. Potentially, the absence of gravity could be advantageous, allowing for more creative and flexible positioning systems. It may be that an entirely new approach to positioning (for example, movement from Cartesian to polar coordinate systems) for part production is needed, with linear motion systems being replaced by rotary (harmonic drive) systems. Robotic interaction with the additive manufacturing process may be required in the absence of gravity to position and constrain the geometry of the developing part.

At the same time, the lack of gravity presents possibilities for additive manufacturing in space not available to ground-based machines. The lack of gravity might permit a printer to work on the "bottom" and the "top" of an object at the same time. Imagine a printer for use in space that has multiple print heads and works on all six sides of an object resting in the space between the heads. Air jets or electrostatic attraction might be used to keep the growing object in place, or even to move it to the orientation most suitable for printing. Put another way, instead of thinking of lack of gravity as a constraint or an environmental problem to overcome, it may be possible to think of microgravity as an opportunity to explore entirely new techniques.

### Thermal Environment

The thermal environment of space will pose challenges to any additive manufacturing technique. Thermal effects related to the lack of convection will impact many of the targeted processes, whether the system is internally or externally located. In addition, an externally placed additive manufacturing system operating in Earth orbit will experience similar thermal loads of solar, albedo, and Earth infrared during an orbit, as would a spacecraft. Both the operation and performance of the manufacturing system and the dimensional accuracy of the product being produced will be impacted. For example, maintaining accurate physical dimensions throughout the day/night cycle and subsequent thermal fluctuations will be challenging. However, there are potential solutions, such as shielding the equipment behind a sunshade, as is done for some space-based telescopes.

### Quality, Verification, Validation, and Functional Testing

On-orbit manufacture of hardware will require techniques for part quality assurance, process verification, and functional validation that can be executed in space, either with or without human intervention. This part-certification process has to be verified in space where adverse conditions of remoteness and visual impairment due to unique white-light conditions have to be taken into account. It is likely that autonomous inspection of the manufacturing process in situ will be a promising methodology. In situ monitoring of flaw detection will help inform decisions about final material properties, part performance, part quality, and recovery on the fly, and in rare cases, the need to scrap the part and start over. Process sensitivity to the space environment, which includes microgravity and thermal conditions, will have to be fully characterized to ensure repeatability and thus part quality. Because the degree to which these various factors influences additive manufacturing is still being investigated in 1 *g*, appropriate measurement and testing will be required for space applications.

A database of effects of objects will have to be developed to calibrate the quality assurance methodology and to set part acceptance criteria, including whether a part is inspected in situ or post production or at some intermediate state, and to establish a non-destructive inspection methodology. Cost-benefits trade study data will have to be produced and analyzed. The level and type of validation and verification required will depend heavily on the complexity of the product being produced. Approaches designed for qualifying simple parts may not be scalable to an additively manufactured complex system. These are some key gaps that need to be filled.

### Infrastructure

Additive manufacturing machines do not operate without numerous ancillary systems. The facility provides shelter from the elements as well as a stable environment, power is delivered, human beings provide means for machine preparation and post processing and qualification of parts as well as clean up and reset of the process for the next production run. When considering the transfer of any additive manufacturing technology to the space environment, some attention will have to be focused on what level of infrastructure will have to be constructed to support the manufacturing capability.

The level of required infrastructure will depend to a great extent on the type of processes used and the design of the equipment. It is not hard to speculate what some of the common denominators may be across the potential techniques. For example, to develop any kind of space-based manufacturing capability, a platform stable enough to meet the limitations of the processing technique, that is, minimizing any external forces due to rotational or vibrational forces, is required. Another important part of the required infrastructure that will have to be created is a power collection, storage, and distribution system. A further requirement will be some level of data and telemetry exchange with a remote control station located on Earth. The need for avionics will drive the introduction of a thermal control system to help regulate the thermal environment that the electronics will be subjected to. Some level of autonomy, above and beyond that utilized in terrestrial processes, will be required for a manufacturing process based in space.

Other considerations, beyond those of infrastructure, will have to be addressed when designing a platform for additive manufacturing in a remote, hostile environment, including manufacturing process parameters such as material handling, pre- and post-processing, quality control, final product disposition, and so on. Ancillary manufacturing activities, which involve human engagement on Earth, have to be fully automated when implemented on an orbiting platform. Trade-offs will have to be made when choosing between an additive manufacturing facility design to host humans versus an autonomous facility requiring no human presence.

### Stable Manufacturing Platform

Many additive manufacturing techniques monitor flow and flow control parameters to control the build process. It is likely that any externally induced rotational or vibrational forces can negatively impact the integrity of the final part. Vibrational loads induced by the activities of crew members, the background "white noise" of operating machinery, or the occasional attitude control adjustment can all affect the integrity of a part manufactured with additive technologies. A vibrational isolation system coupled with a stable platform design will be important for ensuring high-quality manufacturing. The potential need for a stable platform implies some level of dynamic control system, the complexity of which will depend on the level of stability required. The fundamental components of a motion-control system include the stabilization system, gyroscopes, and sensors capable of indicating rate and direction of motion. Dynamic control based on a propulsion-type subsystem may or may not be optional. In addition, the attitude control system must be capable of managing angular momentum during the manufacturing practice to counter a combination of solar, gravity gradient, magnetic, and atmospheric drag effects. Once NASA and the European Space Agency begin conducting experiments on the ISS, they will begin to define the acceptable forces for these processes.

The manufacturing process itself could also potentially create demands on the design and deployment of an attitude control system. During the course of building a system via additive manufacturing, material (mass) is constantly being deposited, creating a dynamically changing geometry. As parts grow in cross section, disturbance torques may change, forcing a control system to compensate. These effects will be pronounced in the development of large cross-sectional areas.

The creation of a stable on-orbit manufacturing platform is required. Extensive experience exists both in the design and engineering of systems that have to survive and operate in a space environment. The main concern with the creation of a platform for locating additive manufacturing in space is determining requirements necessary for ensuring a quality product.

## Communication: Data and Telemetry

Communications will be required for control process, file transfer, quality control and final product analysis, in addition to likely characterization during the manufacturing process. The nature of the type of information that has to be transmitted back and forth (data and video, bandwidth, number of simultaneous transmissions) will drive the complexity of the avionics system and the computer and software control system. Further data exchanges will likely be necessary to monitor the health and status of the manufacturing platform itself. The choice of orbit (altitude and inclination) will drive the requirements for a thermal control system, active or passive, to ensure system functionality across a potentially dynamic thermal environment.

## Power

Not only will additive manufacturing equipment need power for manufacturing operations, but the platform itself will require power to support the manufacturing process and subsystems involved in maintaining the platform. Power systems readily available to operate in a space environment include solar and nuclear. Storage batteries that function in space are also available.

Present day additive manufacturing machines use electric power at various rates, depending on specific deposition processes and the type of heating (i.e., plastic being heated by a laser or metallic powder energized by an electron beam). The deposition power levels vary from several hundred watts for electric heaters and lasers working with polymer materials to several kilowatts for electron beam systems. However, the total energy and build time for additive manufacturing of an artifact depends on the time required to heat and deposit the requisite amount of material as well as the overall efficiency of the machine. Using a relative measure of power—the number of kilowatt hours per kilogram (kWh/kg) of deposited material—can highlight the challenges or compatibility of different additive manufacturing processes for use in space.

Three studies of the energetics of current commercial additive manufacturing machines have been published over the past 15 years for different additive manufacturing lay-down processes. The results show energy consumption rates ranging from 17 kWh/kg (electron beam melting of vanadium alloy) to more than 40 kWh/kg for laser sintering. The wide range of results is typical of the varying efficiencies of the melting and deposition of the build stock, as well as the mechanical dissipation of complex machinery needed to accurately move the deposition head across the bed of the device. According to Wohlers,[3] there are presently 14 global and 23 regional additive manufacturing equipment manufacturers producing hundreds of different machines. In addition, major aerospace and other companies have developed their own additive manufacturing devices. All of these have differing energy profiles.

Table 3.1 shows the differences between different additive manufacturing processes in terms of materials, typical system power, and energy consumption rates.

[3] T.T. Wohlers, *Wohlers Report 2014, 3D Printing and Additive Manufacturing State of the Industry*, Annual Worldwide Progress Report, Wohlers Associates, Inc., Fort Collins, Colo., 2014.

TABLE 3.1 Energy Consumption of Five Additive Manufacturing Materials

| Process | Material | Average Power (kW) | Energy Consumption (kWh/kg) |
|---|---|---|---|
| Stereolithography | Photopolymer | 0.88 | 20.7-41.4 |
| Selective laser sintering | Polyamide and typically other semi-crystalline thermoplastic polymers | 0.3 | 29.8-40.1 |
| Fused deposition modelling | ABS, PLA, polycarbonate and typically other amorphous thermoplastic polymers | 1.3 | 23.1-346 |
| Selective laser melting | Stainless steel, SAE 316 L | 0.4 | 31 |
| Electron beam melting | Titanium, Ti-6Al-4V | 3 | 17 |

NOTE: ABS, acrylonitrile butadiene styrene; PLA, poly(lactic acid) or polylactide; SAE, SAE International.
SOURCE: Data from M. Baumers, C. Tuck, R. Hague, R. Wildman, and I. Ashcroft, A comparative study of metallic additive manufacturing power consumption, pp. 278-288 in *Proceedings of the Solid Freeform Fabrication Symposium 2010*, University of Texas, Austin, 2010, http://utwired.engr.utexas.edu/lff/symposium/proceedingsArchive/toc.cfm; Y. Luo, Z. Ji, M.C. Leu, and R. Caudill, Environmental performance analysis of solid freeform fabrication processes, in *Proceedings of the 1999 IEEE International Symposium on Electronics and the Environment*, Institute of Electrical and Electronics Engineers, 1999, http://ieeexplore.ieee.org/, doi:10.1109/ISEE.1999.765837; R. Sreenivasan and D.L. Bourell, Sustainability study in selective laser sintering—An energy perspective, in *Proceedings of the 2009 Solid Freeform Fabrication Symposium*, University of Texas, Austin, 2009, http://utwired.engr.utexas.edu/lff/symposium/proceedingsArchive/toc.cfm.

These data are important in assessing the ability of a remote facility to construct various types of space hardware, including complete satellites. The power needed by the additive manufacturing machines listed in Table 3.1 are within the expected norms of ground-based facilities, which are able to draw significant electric power from terrestrial grids. For example, 6.6 kW (30 A of 220 V a.c.) power is readily available in any industrial facility, 7 days per week, 24 hours per day. In space, electric power is an expensive commodity, most often derived from arrays of photoelectric cells exposed to solar radiation. In space, the power requirements of several additive manufacturing machines each averaging 3 kW, plus supporting local infrastructure, may place significant strains on overall platform designs using currently available, commercial solar power arrays.

State-of-the-art solar arrays are available with output powers up to about 25 kW, along with large battery systems needed for periods of orbital darkness. This latter requirement depends on the specific orbit, both altitude and inclination, of the manufacturing platform. The design of an appropriate-sized power generation and storage system will, in turn, drive the mass of the platform and influence other subsystem design. Of course, the choice of orbit will also have major operational impacts that will have to be considered in the cost-benefit equation. For example, if the completed product has to be transported to a different orbit, that will require fuel and a transfer spacecraft.

## AUTONOMY

The level of autonomy that is necessary in a space-based manufacturing platform designed to build a satellite using additive manufacturing techniques will obviously depend heavily on the operational concept derived for the factory. This raises important considerations.

### Use of a Platform

A platform required for only a unique, one-time use will have different design drivers than a platform that will act as a base for a longer-term, space-based manufacturing capability. A one-time-use system will likely have a fairly integrated subsystem design across the required infrastructure and the manufacturing-related subsystems. In this case, for example, the raw materials can be designed into the system, providing a more "turn-key" manufacturing process. The appropriate characterization, analysis, and quality control sensors have to be designed and built as part of the manufacturing equipment. Any desired post-processing steps will also have to be built into the equipment as well, yet another factor in determining the complexity of the design. At the completion of the project, the product will have to be disengaged from the platform, deployed, and the platform disposed of. (Space debris is fast becoming a major issue, and it is likely that any evolving space manufacturing systems will have to consider their disposal plans as an important factor in their designs.)

A single-use facility, designed for a custom, one-time construction project, will be more complex in design than a general-purpose facility targeted at establishing generic additive-manufacturing capability in space. The exact level of complexity in any platform will be highly dependent on the desired complexity in the final product. The level of autonomy required for a single-use facility is quite high and will further impact the design complexity because the manufacturing platform will have to be designed to launch, commence operations, and deploy with no direct intervention by humans.

If the operational goal is to build an on-orbit facility capable of manufacturing either multiples of one type of satellite or a variety of different types of satellites, the complexity in design and operations increases. A multiuse facility will have different requirements to consider than a single-use platform. Conceivably a multiuse facility will be desired to provide a more "plug and play" environment and will be designed as such. Hence, the platform infrastructure can be designed as separate and distinct subsystems and therefore support a wide variety of projects, both from a material and functional aspect. A crucial piece of infrastructure needed in a multiuse platform, however, is a docking system or some method of delivering raw materials and material related to maintenance and upkeep of the facility. The exact design and level of autonomy required in a multiuse platform will be determined by the operational concept regarding potential human interaction.

## Human Presence

Additive manufacturing, as it is practiced today on the ground, depends heavily on the active presence of human operators for system preparation with software loading, system checkout, verification of settings, verification of quality of lay-down construction, verification of metrology, removal of supports (if any), and other features of the desired mechanical and material features of the object under construction. Additively manufactured parts also require post-processing steps that humans currently fulfill as well as cleanup and reset of the equipment for the next production run.

A concept of operations assuming a limited human-in-the-loop requirement will reduce the complexity of autonomy required in a long-term, space-based manufacturing process. Humans can visit the platform between manufacturing runs to manually complete and close out the just-completed process while additionally setting up for the next production. This includes delivering and loading the raw materials, disposing of the waste, final post-processing, testing, or quality check of the manufactured craft, and even deployment. Even just a limited human-tended capability will simplify the platform design and minimize the required automation. The inclusion of humans as part of the process, although simplifying the design and production of the platform, will add complexity to the operations, however. Considerations such as transportation and the working environment are significant, even though solutions can be built on experience and existing knowledge bases.

The lack of human tending will drive requirements for automation in areas related to rendezvous and docking, materials handling, equipment set-up, final processing or product deployment, and waste handling, among other things. The complexity of the automation systems will be largely determined by the required flexibility and the nature of the processes and products. Investments in human telepresence and robotics, at a minimum, will be required. NASA is already conducting research on both human telepresence and robotics on the ISS. For example, the Robonaut program and a telepresence experiment conducted on the International Space Station (ISS) in 2013 are examples of technology that can potentially have applications to in-orbit manufacturing and assembly.

The use of multiple material components typical of all satellite systems could impose continual inspection and verification requirements during and after additive manufacturing production processes—in other words, additive manufacturing may require much greater monitoring during production than other forms of production. As previously mentioned, verification of physical and electrical continuity of electric power and digital communication links are critical in establishing the quality of the final product. Furthermore, present-day satellites require fast, radiation-resistant digital processors and dedicated chips. Considerable engineering effort will be required to integrate and test electrical components produced with additive manufacturing. Finally, the state of the art of other operational equipment, such as optical, infrared, and radio sensors, is of such sophistication as to require not only very careful preflight verification and calibration, but also needs in-flight testing capabilities that are likely to challenge additive manufacturing processes with respect to materials, resolution, radiation hardness, and other features for many years.

**Finding:** Autonomously meeting all of the requirements can be daunting; even on Earth, no fully autonomous manufacturing and verification process for producing a satellite exists.

**Finding**: Production of additive manufacturing components on the ground currently requires extensive human presence and participation. Automated manufacturing capabilities on the ground are currently under development. However, significant further development will be required for automated space-based additive manufacturing, and much of this development is likely to require government support.

## CHALLENGES RELATED TO ADDITIVE MANUFACTURING ON THE INTERNATIONAL SPACE STATION

The ISS provides a convenient and natural platform for the evolution of additive manufacturing to a space-based environment (Figure 3.4). Not only does the ISS provide a place to study the effects of the unique aspects of the space environment on additive manufacturing (microgravity, thermal environment, etc.), but it also is a potential customer of additive manufacturing, and its ability to create parts on demand for maintenance and repairs and thus can provide immediate technology demonstration and operational impacts. There are several steps in the

FIGURE 3.4 Microgravity glovebox on the International Space Station (ISS). The ISS has equipment that can be used for additive manufacturing experiments. Any such experiments will naturally have to compete with other research priorities. SOURCE: Courtesy of NASA.

evolution of the additive manufacturing capability that can be tested on the ISS, both in the basic science area as well as examining potential engineering adaptations of the processing equipment itself.

### Internal

Adapting additive manufacturing for operations in the internal volume of the ISS requires attention to several issues not present for ground-based processing. Besides the examination of the effects of microgravity, some engineering challenges related to equipment design and operation in a human-tended closed environment will need to be addressed. NASA has specific requirements that spaceflight hardware has to comply with in order to be compatible, not only with the environment and resources available on the ISS but also safety-related areas that deal with the human presence. While none of these are insurmountable, they impose an added layer of complexity on the design of an additive manufacturing system for placement on the ISS.

Initial investigations in a location that can be tended by humans are very useful, however, because there are several aspects of additive manufacturing production methodologies that require human interaction. At this time, the loading of the feedstock and the setup, preprocessing, and any required post-processing of the parts all require human intervention. In the initial investigative and development stages, having humans in the loop to assist in the characterization and quality of the final part is also very helpful.

### External

The exterior of the ISS is available to perform experiments and technology demonstrations in the natural environment of space and is a natural place to conduct the research leading toward an independent space-based manufacturing capability (Figures 3.5 and 3.6). Moving the capability external to the ISS, however, introduces some additional basic technology challenges. One of the main technical issues with additive manufacturing is induced thermal stress during the processing. Placing the processing in the extreme thermal environment in space will only exacerbate thermal issues. Some basic science and engineering studies will need to be done to increase understanding of the process and the effect of the vacuum of space and a thermally dynamic environment on finished parts. Likely some engineering design alterations to the manufacturing equipment itself will also have to be considered. There is already a lack of information related to material properties, and the added complexity of a dynamic thermal environment will create further complications. It is likely that additional research and parametric studies will be required beyond those needed on Earth in order to achieve a clear understanding of what additive manufacturing in space can achieve.

One major step that will occur as the process is moved externally on the ISS is a decrease in the availability of human interaction with the equipment during the various stages of processing. The level of autonomy of the equipment will have to increase. A human-tended capability via spacewalks or robotic interface will be available, but by definition, it will be much less flexible and available as compared to an internally based system. The preparation of the feedstock can be done internally by humans and readied for insertion into the machine. Ideally, the equipment could be designed so that humans could insert the feedstock, again either manually or robotically, and retrieve the final sample for post-processing and analysis (or a return to Earth where this can be done). The system, however, would have to potentially become more autonomous than an internal system and would require self-sufficiency for the complete process, any material changes (in the case of a multi-material system), or handling requirements. This includes the disposition and management of waste material, thermal management techniques that are determined to be necessary not only for the process but also for the equipment components, in situ observation and recording, and data handling and transmission.

The experience gained from the development of an additive manufacturing technique for use on an external platform on the ISS will be directly applicable toward the development of a fully autonomous stand-alone, free-flyer-based capability, one that could potentially be used to manufacture a spacecraft on orbit.

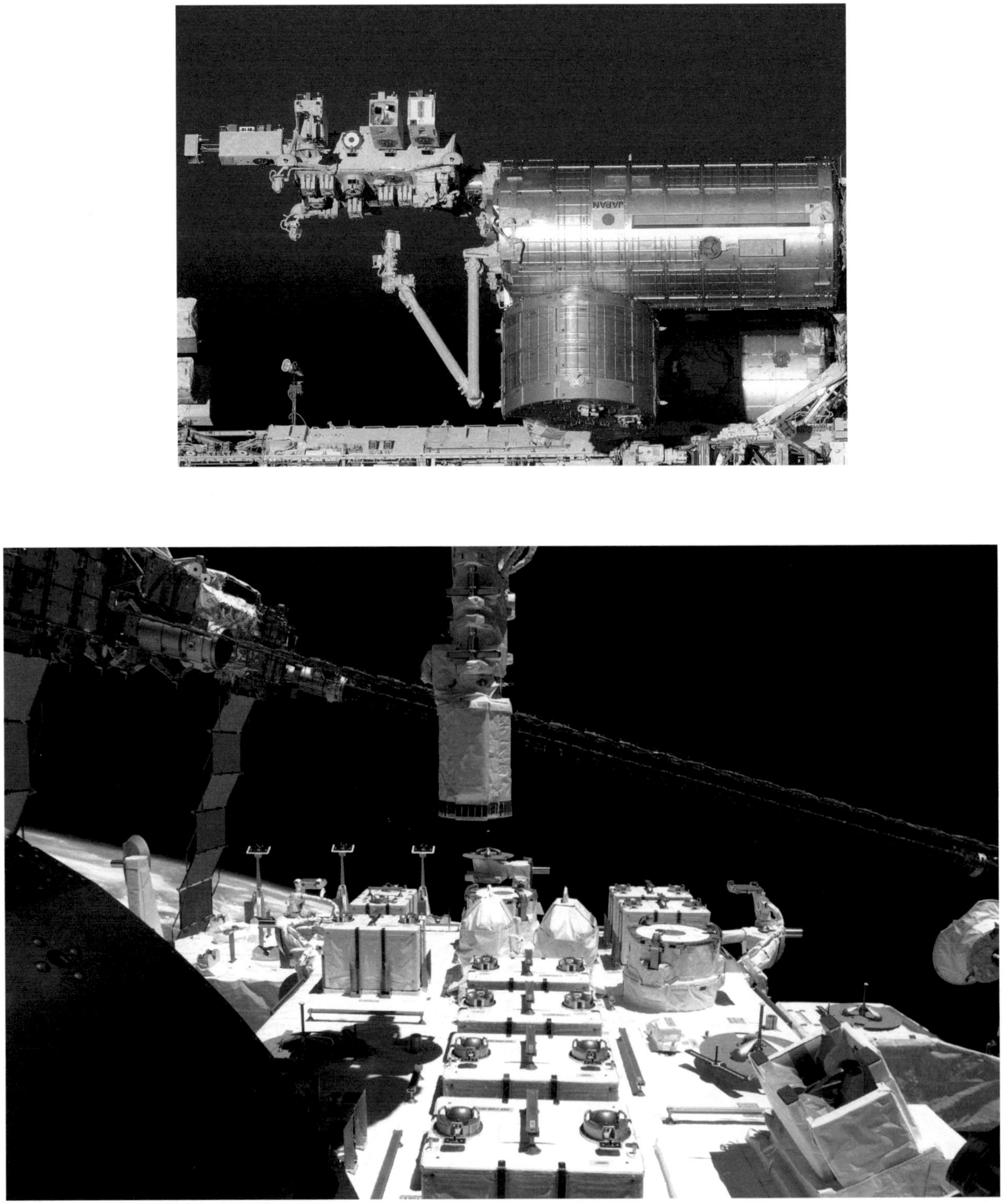

FIGURE 3.5 The Japanese Experiment Module, also known by the nickname Kibo, features an exposed facility that can serve as an experiment platform for additive manufacturing. SOURCE: Courtesy of NASA.

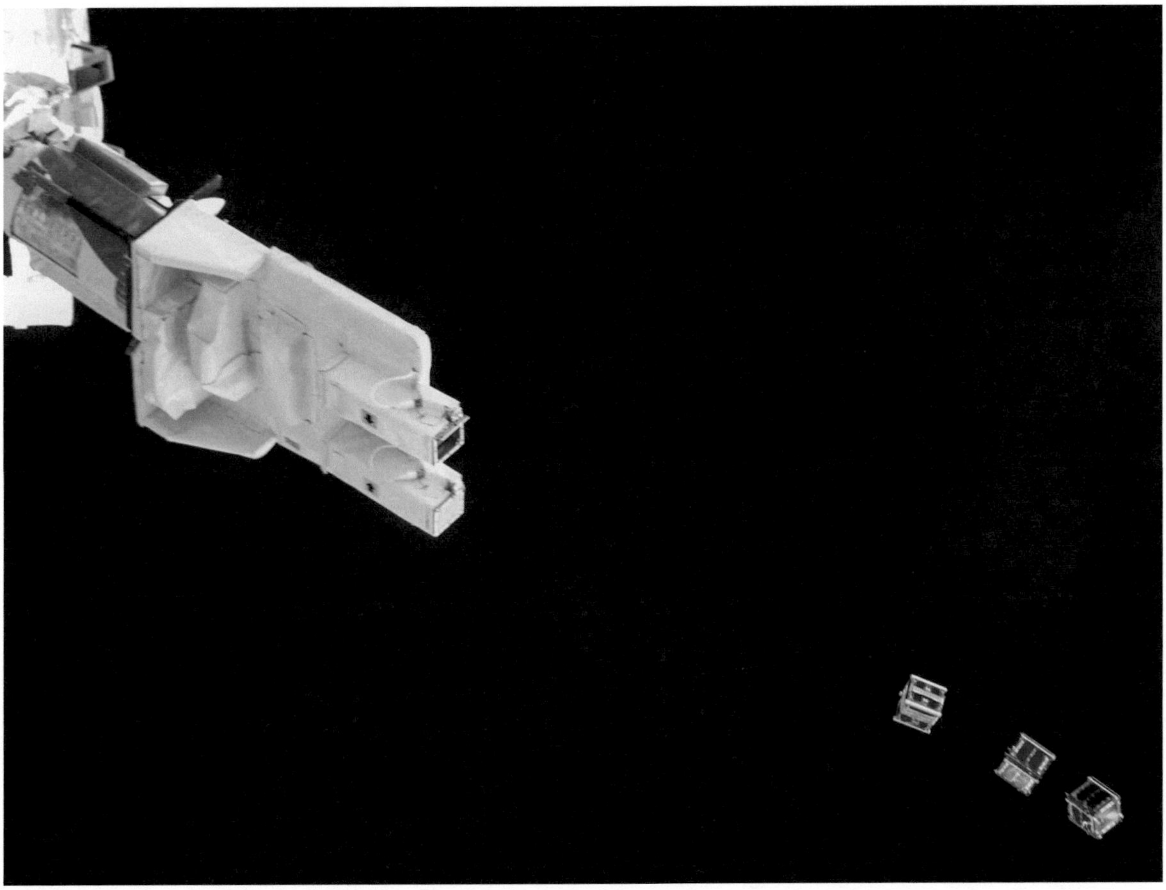

FIGURE 3.6 CubeSats being deployed from the International Space Station after being removed via the airlock in the Kibo module. SOURCE: Courtesy of NASA.

## ADDITIONAL CHALLENGES RELATED TO FREE-FLYER PLATFORMS

As mentioned earlier, there are several steps that additive manufacturing evolution needs to complete terrestrially before moving to space. Being able to integrally build a system with multiple material systems and integrated capabilities is a critical technology development that terrestrial additive manufacturing needs to advance through. Once the many terrestrial-based problems have been solved, and after the subsequent completion of a development program that adapts additive manufacturing techniques to an external platform on the ISS, the fundamental technology, process parameters, and basic equipment design required to build a space-based manufacturing capability around additive manufacturing should be well understood and demonstrated. The level of further development that is required to fully use additive manufacturing to build a spacecraft on orbit is more related to the required autonomy necessary to support the complete manufacturing process. Even though various levels of complexity can be foreseen, depending on the exact spacecraft specifications, the manufacture of any satellite in space requires certain support equipment and fundamental resources. On Earth, this infrastructure exists as part of the normal background. In space, all necessary resources will have to be delivered and built to support a manufacturing capability. In addition, the need for regular maintenance and repair of the space-based manufacturing facility is also an important aspect to take into account.

There are several free-flying spacecraft that are either currently available or will probably become available within the next decade that could serve as a point of entry for free-flying test beds for additive manufacturing in space. These include SpaceX's Dragon and its planned free-flying Dragon Lab, Orbital Science's Cygnus, and Bigelow Aerospace's inflatable structures (Figure 3.7). In addition, Sierra Nevada is currently developing its Dream Chaser space plane, and the U.S. Air Force operates the Boeing X-37B space plane. These vehicles offer small volumes, both pressurized and unpressurized, which could serve as free-flying platforms for experiments targeted at understanding the requirements for developing a sophisticated additive manufacturing capability in space. Several of these vehicles can also provide a human-tended capability, outside of the ISS, that would allow independent operations and the capability to dock or re-dock experiments for retrieval.

Although these platforms have potential for research use, they also have limitations that would preclude them being used for a permanent manufacturing facility. While these vehicles do have dynamic control systems, communications, and telemetry capability, the available volume is a concern—their internal volumes are quite small. The size of a manufacturing facility in orbit is dependent on the scale of the products for manufacture. Whether a small vehicle free flyer or a Skylab-sized orbiting facility is necessary will be driven by satellite requirements. In addition, power needs will be a major factor, as will the level of automation needed to conduct the manufacturing process. It might be possible to design free-flying experiments that could operate out of the space station, but that will add complexity to the ISS missions. In any event, the design, construction, and operation of any free-flyer platform for the support of an established manufacturing capability on orbit is a major undertaking.

## ADDITIONAL CHALLENGES RELATED TO IN SITU-BASED PLATFORMS

Many of the technological hurdles associated with moving additive manufacturing from a terrestrial environment to a space-based environment will also facilitate the adaptation of this technology to other space uses. A natural use of additive manufacturing techniques for the purposes of human exploration is for in situ resource utilization on planetary bodies where humans want to establish a presence. This will present additional challenges, as the goal will be to use resources on the planetary body in additive manufacturing machines, and their suitability may be difficult to determine. The slow build up of technology, experience, and knowledge gained from adapting and developing additive manufacturing to on-orbit operations provides a solid foundation for developing equipment and techniques for use elsewhere. The evolution of the infrastructure systems and platforms needed to support additive manufacturing equipment and processes, the level of autonomy developed to manage on-orbit untended operations, and the creativity in designing additive manufacturing devices that can build something bigger than they are will all facilitate in situ resource technologies and techniques.

## SUMMARY

Additive manufacturing in space is much more of a systems engineering and industrial logistics problem compared to additive manufacturing on the ground. In addition to the constraints imposed by the space environment, issues such as supply-chain logistics, integrated processes, minimal human interaction, and quality control are more pronounced. Supporting infrastructure and environment, which are relatively straightforward and easy considerations on the ground (i.e., rent factory space, connect to the local power grid), are not simple for space. Nevertheless, NASA has already taken major steps to develop a technology and infrastructure base that can support at least initial experimentation.

**Finding:** Additive manufacturing techniques and processes require supporting infrastructure in order to be successful. This infrastructure includes a reliance on human intervention for important steps to achieve a final product. Transplanting an additive manufacturing capability to space requires consideration of how the supporting infrastructure, including the applicability or desirability of maintaining humans in the loop, needs to be evolved to operate in the new environment.

**Recommendation: Actual costs of the reproduction of components or spacecraft should not be the sole criterion for evaluation of the benefits of additive manufacturing; criteria should also include the value of creating structures and functionalities not feasible before.**

FIGURE 3.7 Cygnus (*top*) and Dragon (*bottom*) are two spacecraft that could be used as free-flying research platforms for additive manufacturing experiments. SOURCE: Courtesy of NASA.

**Recommendation: As the technology evolves and as NASA and the Air Force consider projects utilizing this technology, they should jointly undertake a cost-benefit analysis of the role of space-based additive manufacturing in the construction of smaller, more reliable, less massive satellite systems or their key components.**

Such an analysis will not be simple, nor is it likely to be complete. Based on evidence that a basic analysis along these lines was already attempted within the Air Force, the committee believes that further coordinated work could potentially help clarify research funding decisions.

**Recommendation: When considering moving additive manufacturing technology to the space environment, any person or organization developing plans should include in their planning the infrastructure required to enable fabrication processes based on additive-manufacturing, such as power, robotics, and even human presence. Studies examining the types of infrastructure should be undertaken in tandem with the development of the additive manufacturing technology itself.**

# 4

# A Possible Roadmap for NASA

NASA's mission directorates that have primary interest and/or applications for space-based additive manufacturing are the Space Technology Mission Directorate (STMD), the Human Exploration and Operations Mission Directorate (HEOMD), and the Science Mission Directorate (SMD). STMD has primary responsibility for NASA space technology insertion, including small satellites, and is the point of contact for the NASA element of this study. STMD is necessarily the primary and initial NASA stakeholder in development and application of space-based additive manufacturing. The directorate has begun efforts to identify and prioritize an initial roadmap for advanced manufacturing in general, which specifically includes elements of a future space-based additive manufacturing strategic plan and implementation strategy. The committee believes that this is a good first step.

SMD is responsible for developing its own mission-specific technology. However, SMD often also looks to the technology development efforts of STMD and HEOMD to provide technology products and guidance for potential inclusion in their principal investigator-led investigations. SMD is primarily interested in robotic space exploration and operates a large fleet of satellites to conduct Earth science, astronomy and astrophysics, heliophysics, and planetary science research missions (Figure 4.1). Many of these missions will undoubtedly benefit from ground-based additive manufacturing, and in fact the first additively manufactured parts to be flown in an SMD space mission are aboard the Juno spacecraft that will enter Jupiter orbit in 2016. Planetary science research could possibly utilize space-based additive manufacturing for future missions involving lunar, asteroid, or martian surface operations. However, SMD will most probably not be an early user or directly involved in space-based additive manufacturing activities, other than as a possible recipient of products developed that can satisfy their science and mission requirements.

The International Space Station (ISS), managed by HEOMD, is the prime candidate for advanced concept technology demonstrations of space-based additive manufacturing technologies, processes, and products, as well as an initial test bed and staging platform for utilization and optimization of products developed with space-based additive manufacturing. In addition to ISS space hardware repair, replacement, and repurposing and manufacturing of essential experiment-unique hardware, space-based additive manufacturing products could also be developed for applications such as in situ resource utilization, life support, synthetic-biology-based biomanufacturing, disposable medical products and devices, food production, and astronaut-specific interface items. In the near-term (less than 5 years), the possibility exists to produce small satellite systems such as CubeSats on the ISS, using a combination of standardized functional components, transported from the ground to the ISS, coupled with space-based additively manufactured structures, interfaces, and payload elements.

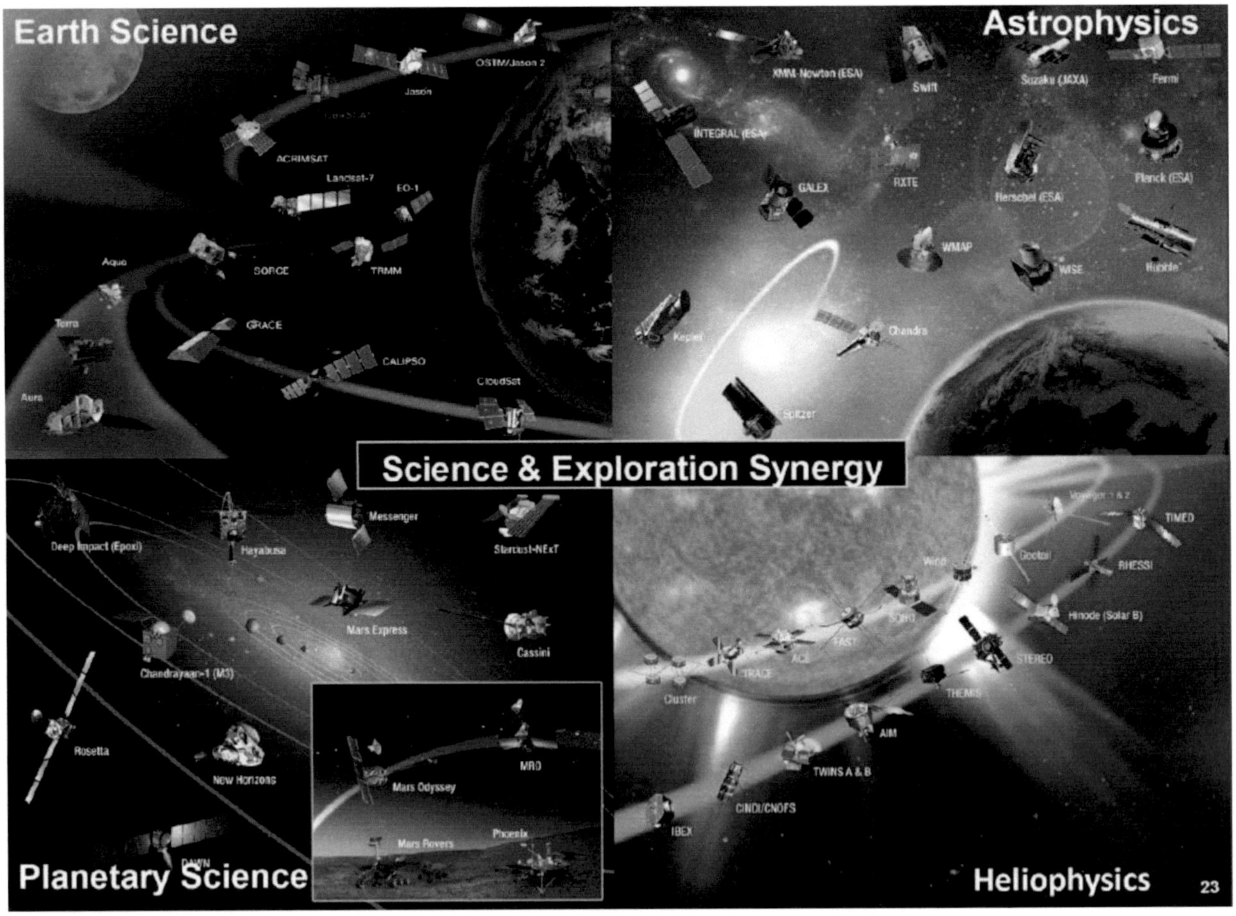


FIGURE 4.1 NASA's Science Mission Directorate operates a large fleet of scientific missions conducting various types of research and participates in several international missions as well. SOURCE: Courtesy of NASA.

NASA has considerable efforts and activities involving CubeSats and nanosats and has had several successful CubeSat missions over the past 10 years. The first domestic NASA CubeSat launched on a U.S. vehicle was GeneSat, a 3U CubeSat carrying a biological payload, developed by NASA Ames Research Center in collaboration with Stanford University, Santa Clara University, and Cal-Poly San Luis Obispo. GeneSat was launched as a secondary payload aboard the Department of Defense (DOD) TacSat-2 mission on December 16, 2006. There have been several other successful ad hoc CubeSat missions up through the present, and several centers, including Ames, the Jet Propulsion Laboratory (JPL), Goddard Space Flight Center (GSFC), and Johnson Space Center, have initiated or flown CubeSat/nanosat missions and technology demonstrations. Most recently, NASA Ames Research Center, in collaboration with San Jose State University students, successfully launched the first CubeSat from the ISS, TechEdSat. Other centers, particularly JPL and GSFC, have begun to place particular interest on use and application of smaller satellites and nanosatellites, for both early stage technology demonstration, as well as to implement science missions, possibly including future interplanetary and deep-space nanosatellites.

In 2011, STMD created the Small Spacecraft Technology Program (SSTP), whose objectives are the following: (1) identify and support the development of new subsystem technologies to enhance or expand the capabilities of small spacecraft; (2) support flight demonstrations of new technologies, capabilities, and applications for small spacecraft; and (3) use small spacecraft as platforms for testing and demonstrating technologies and capabilities that might have more general applications in larger-scale spacecraft and systems. As stated by the program, "All efforts focus on small spacecraft capabilities that are relevant to NASA's missions in science, exploration, space operations and aeronautics including those with crosscutting applications for NASA and other users." Because SSTP is incubating and demonstrating newer, higher-risk technologies, mission directorates and program managers who necessarily have to be somewhat risk adverse, often adopt a wait and see attitude, which hampers communication and collaboration and compromises adoption and utilization of SSTP (and other advanced technology programs) developed technologies. Additionally, because the focus of SSTP is on development of the satellite subsystems and integrated technologies from the platform standpoint, emphasis and focus on specific enabling technology areas necessary for additive manufacturing do not get specific and focused visibility.

As the use and application of CubeSats/nanosats matures and expands from university aerospace education and technology demonstration efforts to peer-reviewed science and technology and specific mission applications, their utility, awareness, and acceptance as viable platforms is becoming increasingly evident in both NASA, DOD, and other government agencies, as well as commercial and entrepreneurial space markets. As that happens, concerns for quality control, mission development efficiencies, and cost reductions become increasingly relevant. A major benefit of using space-based additive manufacturing to build CubeSats or nanosats, compared to building space hardware on the ground, is the relaxation of launch load and environmental stress requirements, as well as the possibility for rapid assembly, integration, and deployment. Additive manufacturing is unlikely to reduce the mass of these already lightweight satellites, but the structure design may not have to consider launch loads and vibration.

**Finding:** For additive manufacturing in space, considering a 20-year time horizon, NASA has a unique opportunity to encourage innovative thinking about how to capitalize on the lack of gravity or the lack of atmosphere to achieve

- More rapid formation of objects similar to those made on Earth and
- Objects better than those which can currently be produced on Earth.

## EVOLUTION OF NASA ADDITIVE MANUFACTURING ACTIVITIES ON EARTH AND IN SPACE

There is already activity on the part of NASA field centers and their contractors in the field of engineering development of additive manufacturing. At the agency and mission directorate level, additive manufacturing efforts are managed from STMD. The Advanced Manufacturing Strategic Technology Development Project, which involves multiple centers and discipline areas, represents NASA on the National Advanced Manufacturing Initiative Committee. STMD involvement and interests cross all technology readiness levels (TRLs), from low-TRL activities, including the Materials Genome Initiative, the NASA Institute for Advanced Concepts (NIAC), research fellows, and Small Business Independent Research projects, to higher-TRL technology development and demonstration projects. Examples include the NIAC Printed Electronics Project at JPL, the SSTP Printable Spacecraft Project by COSMIAC at the University of New Mexico in collaboration with University of Texas, El Paso (UTEP), and the Made In Space Technology Demonstration project discussed earlier in this report. For the additive manufacturing of metals, technology development and demonstration efforts are being conducted primarily at Marshall Space Flight Center (MSFC), Langley Research Center, and Glenn Research Center. The NASA Additive Manufacturing Working Group consists of participants from the engineering and technology services and products from all centers. Despite the existence of this working group, the committee learned that additive manufacturing researchers at different centers were not fully aware of work going on at other centers and determined that better agency coordination and communication is needed.

In addition to the agency-level activities described above, there exist additive manufacturing technology projects, expertise, and capabilities at all NASA centers, sometimes as part of engineering and technology organizations, but also in science and technology research and development groups. For instance, NASA Ames recently created the "Space Shop," which is an advanced additive manufacturing facility modeled after the "fab lab" concept (discussed in Chapter 2), which was created at the Massachusetts Institute of Technology Center for Bits and Atoms and co-located with the traditional machine and manufacturing shop. The Space Shop facility is made available as a mentored resource to all who have properly trained on the use and operation of the equipment. Other centers have initiated or have plans for similar, in-house, fab lab-type facilities.

Beyond the engineering and technology activities, and owing to the do it yourself and "Maker Community" visibility, some NASA scientists, principal investigators, researchers, and contractors are also starting to experiment with using additive manufacturing for advanced concept and prototype development of unique and mission-specific hardware and components and as part of their scientific instruments and future space payloads. There have been many early-stage projects, such as development of additive manufacturing technology for synthetic biology-based hybrids, nanotechnology-based additive manufacturing projects, CubeSat-based technology projects, and other science- and instrument-based hardware design and development efforts. These efforts cross virtually all NASA science and technology disciplines and applications.

As often happens in emerging technology disciplines, and particularly in this nascent field of space-based additive manufacturing for space technology applications, the committee has observed an apparent lack of fully coordinated efforts outside of STMD. This could result in disconnects between technology possibilities and applications and additive manufacturing and end-user applications. Some of this is possibly due to the emergent nature of additive manufacturing technologies. In other cases, there is simply no overarching mechanism to facilitate communication and collaboration on additive manufacturing technologies and applications, including space-based additive manufacturing. Such a capability would enable and enhance the development, use, and application of both space- and ground-based additive manufacturing technologies agency wide.

**Finding:** NASA would benefit from coordination of its many and diverse additive manufacturing activities. NASA's full use and application of additive manufacturing technologies, both in space and on the ground, could be made more efficient and effective if there was a stronger associative link between additive manufacturing technology and facility developers and users who may benefit in areas of efficiency, complexity, and cost reductions.

## FACTORS AFFECTING THE USE OF ADDITIVE MANUFACTURING FOR NASA SPACE MISSIONS

Previous chapters of this report discuss the many different additive manufacturing techniques, methodologies, and capabilities now being undertaken worldwide. There are four fundamental factors that will likely have the strongest influence on the future use of additive manufacturing in NASA space applications. These factors can be summarized as follows:

- The degree to which additive manufacturing will provide technical and programmatic benefits. These include reduced mass structures, volumetric efficiencies, increased flexibility in the design-for-space systems, as well as cost and schedule efficiencies;
- The degree to which additive manufacturing production of satellites and other space hardware can be automated and best practices can be shared among NASA centers and contractors;
- The availability of sufficient space-based resources and infrastructure to ensure the effective use of additive manufacturing systems aboard the ISS (e.g., the rate of production of artifacts, the cost effectiveness of the production system, etc.) and in other human exploration missions; and
- The degree to which space-based additive manufacturing technologies and products can be validated as to their utility, efficacy, and applicability and are demonstrated to address and solve specific mission and programmatic needs and requirements.

In order to define and scope a proper space-based additive manufacturing roadmap for NASA to produce useable in-space products, it is not sufficient to describe and consider only additive manufacturing technologies of which, as it has been shown in this report, there are many platforms, processes, and technologies. The environmental conditions and operational constraints in which the space-based additively manufactured product is to be produced will also have to be considered. Accordingly, the roadmap should probably include at least the following three scoping elements:

- The manufacture and production of space-qualified hardware platforms, subsystems, components, and functions necessary to implement the target system;

- The assembly, integration, test and performance verification, and certification test of all elements of the space-based additive manufacturing product; and
- The required infrastructure products necessary for the designated space-based additive manufacturing technique, such as those required for automated or semi-automated sample and feedstock acquisition, preparation, manipulation, and handling of additive manufacturing materials.

Although the demonstration of space-based additive manufacturing onboard the ISS has not yet been accomplished at the time of this report, development of additive manufacturing technology can greatly benefit from human presence.

Each of these factors is important to NASA's use of additive manufacturing. They also help determine the most effective areas of application for additive manufacturing in space activities, up to and including whether or not it is feasible to manufacture a complete spacecraft in space, and if not, what elements can and should be produced.

The committee was impressed with the number of ideas and potential uses for this emerging technology. Although it recognized that some of the ongoing research is proprietary, the committee concluded that there are many people and groups that could benefit from sharing ideas and making contacts while identifying the unique challenges associated with space-based additive manufacturing.

**Recommendation: NASA should consider additional investments in the education and training of both materials scientists with specific expertise in additive manufacturing and spacecraft designers and engineers with deep knowledge of the use and development of additively manufactured systems.**

**Recommendation: NASA should sponsor a space-based additive manufacturing workshop to bring together current experts in the field to share ideas and identify possible research projects in the short term (1-5 years) and medium term (5-10 years).**

**Recommendation: NASA should quickly identify additive manufacturing experiments for all areas of International Space Station (ISS) utilization planning, and identify any additive manufacturing experiments that it can develop and test aboard the ISS during its remaining 10 years of service and determine if they are worthy of flight. NASA currently has methods for providing research grant funding for basic research on additive manufacturing. The agency should closely evaluate them to determine which would allow the most rapid transition of funded research for additive manufacturing to the ISS.**

**Finding:** Because of its broad-reaching activities involving additive manufacturing, NASA could consider establishing or co-sponsoring an ongoing technology interchange forum devoted to additive manufacturing engineering technologies, focusing on serving all NASA centers, universities, small companies, and other organizations. Such a forum could function as a focusing element to orient the agency's efforts and activities in space-based additive manufacturing, providing an integrative, phased capability to identify, facilitate, integrate, and maximize attention and resources to this difficult, long-term objective.

An example of one such forum for a specific technology area is the Small Satellite Conference, the premier conference in this field, which is sponsored by the American Institute of Aeronautics and Astronautics, now in its 28th year of existence. Held annually at the Utah State University, this gathering brings together the small satellite community, including developers, managers, technologists, exhibitors, users, and students from government, industry, and academia, including international participants.[1]

The forum could enable partnerships among NASA, government, university, and industry participants coming together to further develop additive manufacturing, particularly space-based capabilities. The technologies could be developed for applications that enable industry and university participation along with NASA and other government

[1] See, for instance, http://www.smallsat.org.

agencies. With proper support and backing, such a forum could facilitate the leveraging of resources, mechanisms, and necessary infrastructure for efficient coordination and implementation of defined and approved objectives. Such an entity could serve as a resource for communication, collaboration, and interchanges necessary to enable development, integration, validation, and application of required components and subsystems for spaceflight and human and robotic exploration and related terrestrial scenarios.

## ROADMAP CONSIDERATIONS AND CONSTRUCTS

The Space Technology Roadmaps (Figure 4.2) highlight 14 critical technology areas, including those necessary to facilitate robotic human exploration beyond low Earth orbit. The roadmaps target timelines where technology development is needed to enable space exploration, and one of those specifically discusses advanced and additive manufacturing. At the next level, the roadmaps identify specific technology subareas deemed necessary to accomplish the target mission objectives. By definition, NASA, via STMD, and with review and critique from open review and the National Research Council, has approved the format, implementation strategy, teaming approaches, and content and projections put forth in these roadmaps. Figure 4.3 shows an example of a representative roadmap, in this case, for launch propulsion systems.

Beyond the original 14 roadmaps, NASA has begun to identify other technology areas where emphasis is warranted to support and facilitate NASA and national objectives for human and robotic exploration. Although it is not at the same level as the 14 technology roadmaps, NASA would be well served to apply this same approach and strategy in creating an agency-inclusive, overarching, space-based additive manufacturing roadmap. Such a roadmap would help guide the agency and carefully manage its scarce technology development funds.

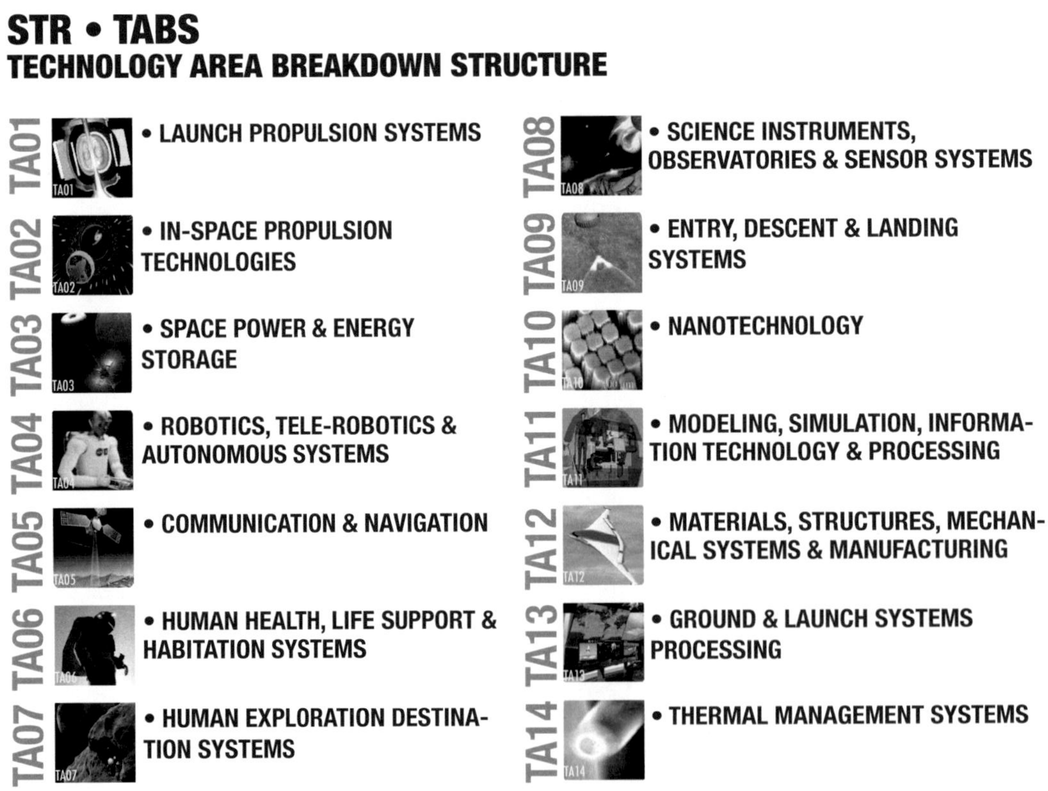


FIGURE 4.2 The 14 current NASA Space Technology Roadmaps. SOURCE: Courtesy of NASA.

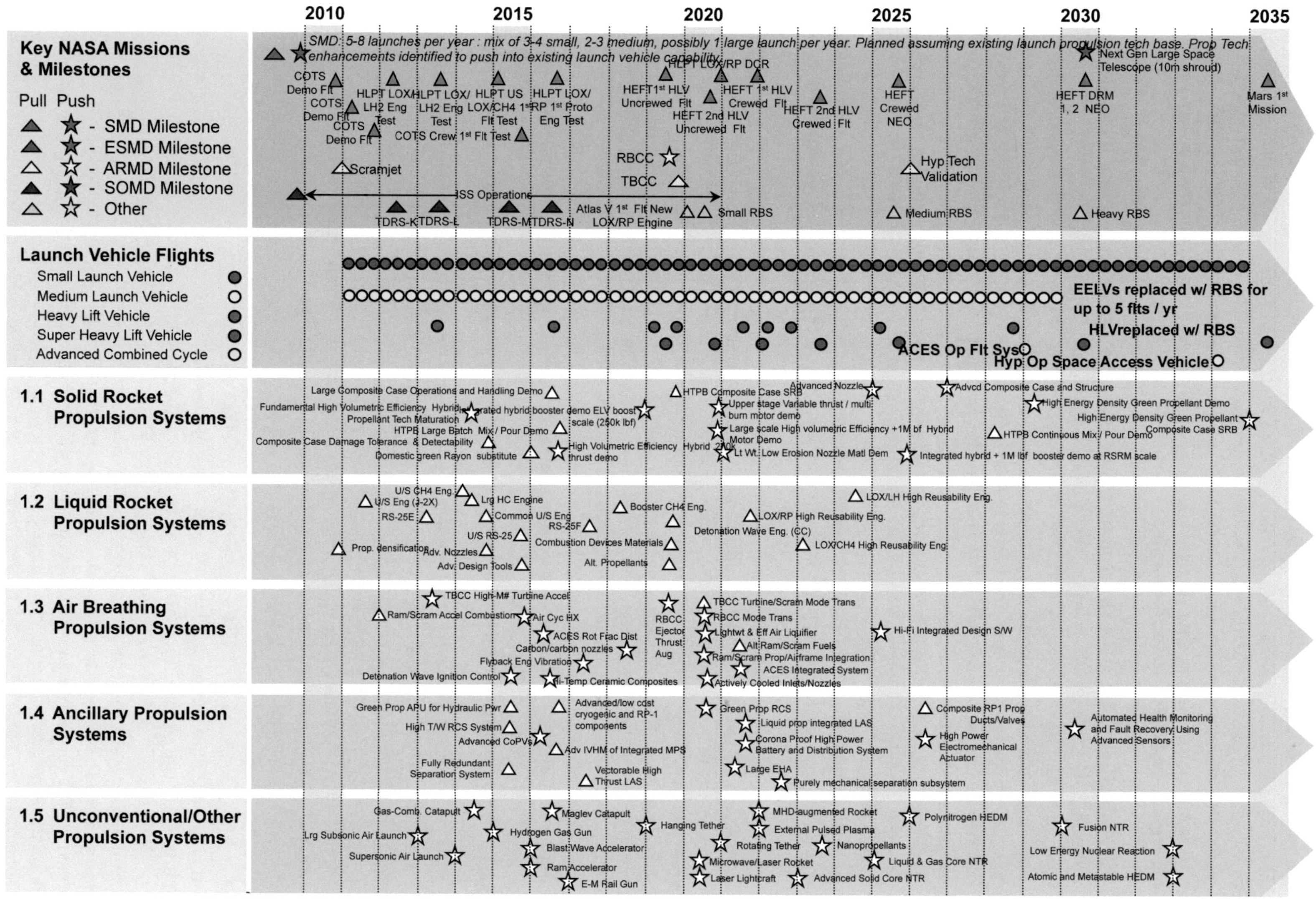


FIGURE 4.3 NASA Launch Propulsion Systems Roadmap. SOURCE: Courtesy of NASA.

**Recommendation: NASA should convene an agency-wide space-based additive manufacturing working group to define and validate an agency-level roadmap, with short- and longer-term goals for evaluating the possible advantages of additive manufacturing in space, and with implications for terrestrial additive manufacturing as well. The roadmap should take into consideration efficiencies in cost and risk management. NASA should build on the considerable experience gained from its Space Technology Roadmaps. The space-based additive manufacturing roadmap objectives should include, but not be limited to the following:**

- **Developing goals for using the technology to assist the agency in meeting its key missions, covering all appropriate mission directorates, especially long-duration human spaceflight and planetary operations, by defining, understanding, evaluating, and prioritizing the direct and supporting technologies for autonomously or minimally attended space-based additive manufacturing and robotic precursor and free-flyer missions;**
- **Identifying flight opportunities, such as on the International Space Station during its next decade of operations;**
- **Targeting the full technology-development life-cycle and insertion strategies through 2050, aligned with target agency missions, for all appropriate mission directorates and related collaborations; and**
- **Ensuring that support for incremental advances to address the technical challenges is supplemented with support for activities related to reaching the full potential of additive manufacturing.**

The incremental advances are likely to be smaller efforts—desktop thought exercises, modeling projects and physical experiments—that invent space-specific additive manufacturing processes rather than adapt current manufacturing methods to the space environment. An example is research related to creating ribbons, trusses, and gossamer arrangements in space. Ribbon structures, for example, require the ability to pultrude thermoplastic ribbons made from carbon fiber. This technique, a combination of material extrusion and sheet lamination techniques, is likely not being explored in any major way in existing federal programs.

**Recommendation: NASA should seek opportunities for cooperation and joint development with other organizations interested in space-based additive manufacturing including the Air Force, the European Space Agency, the Japanese Space Agency, other foreign partners, and commercial firms.**

Figure 4.4 depicts NASA's strategy and capability partitioning for migration from terrestrial additive manufacturing capabilities to utilization of the ISS for both development and demonstration of space-based additive manufacturing, and finally to planetary surface platforms.

MSFC has already taken the lead in developing additive manufacturing in space and has a long history in seeking to develop in-space manufacturing capabilities, sponsoring the first in-space additive manufacturing research in the late 1990s. MSFC has already developed a "Technology Development Vision" that can serve as a basis for an in-space additive manufacturing roadmap (Figure 4.5). However, because agency expertise and talent is spread among the centers, contractors, and universities and research institutions, NASA requires an integrated roadmap developed at the headquarters level and modeled on its previous efforts.

Looking into the out-years at a 20- to 40-year event horizon and the pace of progress that can occur over that period, it is difficult to envision what ground and space-based additive manufacturing capabilities might be achieved by 2040 or 2050. However, given NASA's mission objectives and long-term plans, it should be possible to craft a near-, mid-, and long-term technology development strategy envisioning a convergence of technologies and processes to enable a full-scale, space-based additive manufacturing capability.

Although NASA currently leads in the development of space-based additive manufacturing technology, other organizations also have current or potential interest in developing the technology as well. In the course of developing its roadmap, and certainly following its development, NASA should seek opportunities for cooperation. This

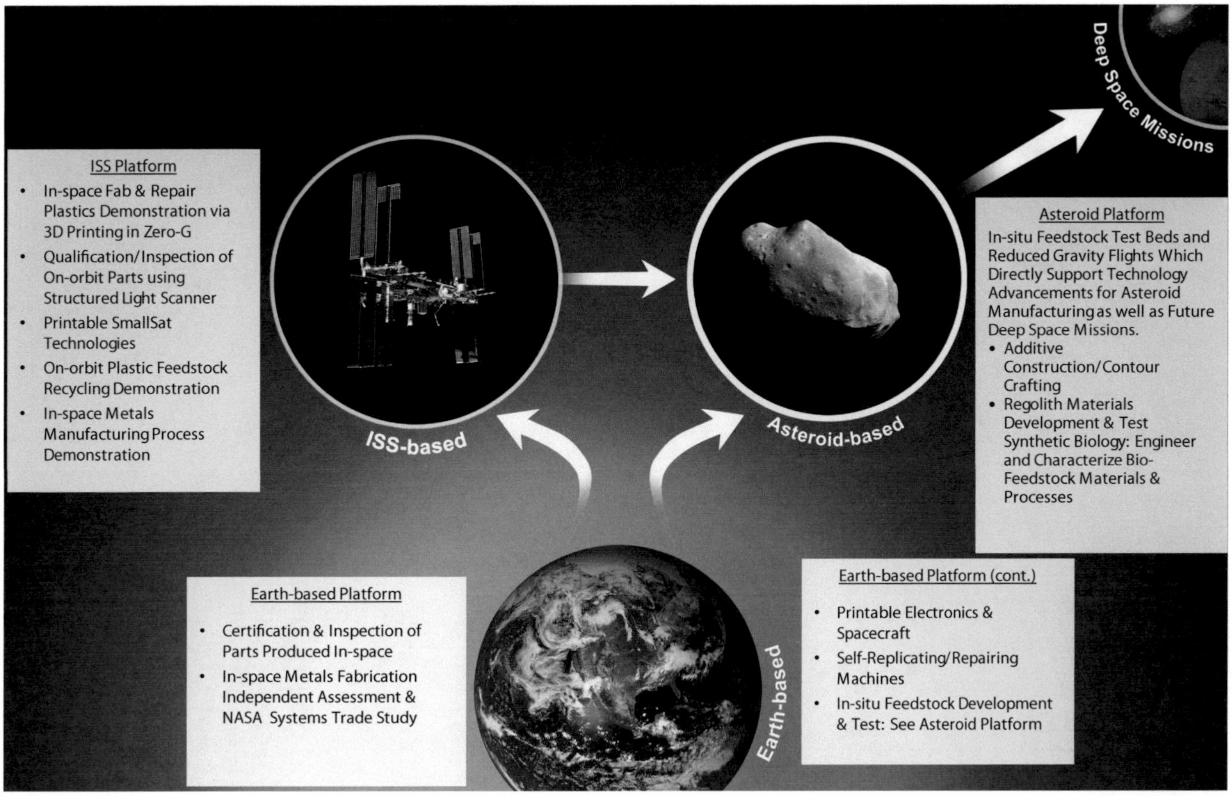


FIGURE 4.4 NASA's general description of available and target platforms for in-space additive manufacturing. SOURCE: Dr. Michael Gazarik, NASA, "Space Technology Mission Directorate Briefing," presentation to the committee, November 12, 2013.

cooperation could take many forms, including NASA making resources available on the ISS in return for access to the results of the research conducted there.

Even as the technology materials, products, systems, and processes are being developed for future in-space advanced manufacturing efforts, and despite the considerable programmatic, technical, operational, and logistical obstacles and hurdles necessary to accomplish these objectives, the committee believes that, in addition to space-based additive manufacturing, the results and products will have considerable benefits for ground-based space systems, platforms, hardware, and product development. These developments will advance the state of the art in automated and autonomous space manufacturing and lead to increased manufacturing and product efficiencies, much as full-scale and semi-attended, robotic-automated manufacturing factories and facilities have benefitted traditional product and manufacturing industries, such as automotive and aerospace and the manufacture of cell phones, personal information devices, and semiconductors.

The committee believes that it is in NASA's interest to continue to define and develop technologies to produce useable in-flight space systems and hardware using both space- and ground-based manufacturing approaches and to define where those activities should occur. In some cases, the options and trade-offs will be based on operational, cost, and logistical concerns. In other cases, it will become increasingly essential to have space-based additive manufacturing capabilities to support NASA's orbital, lunar, Mars, and deep space endeavors beyond low Earth orbit. An iterative, phased approach that evolves from semi- to fully autonomous ground and space production capabilities, to a semi-autonomous (minimally-attended) in-space manufacturing and production factory using a free flyer or platform based on the Moon, Mars, or an asteroid, is probably the best path for NASA if the agency continues developing this technology.

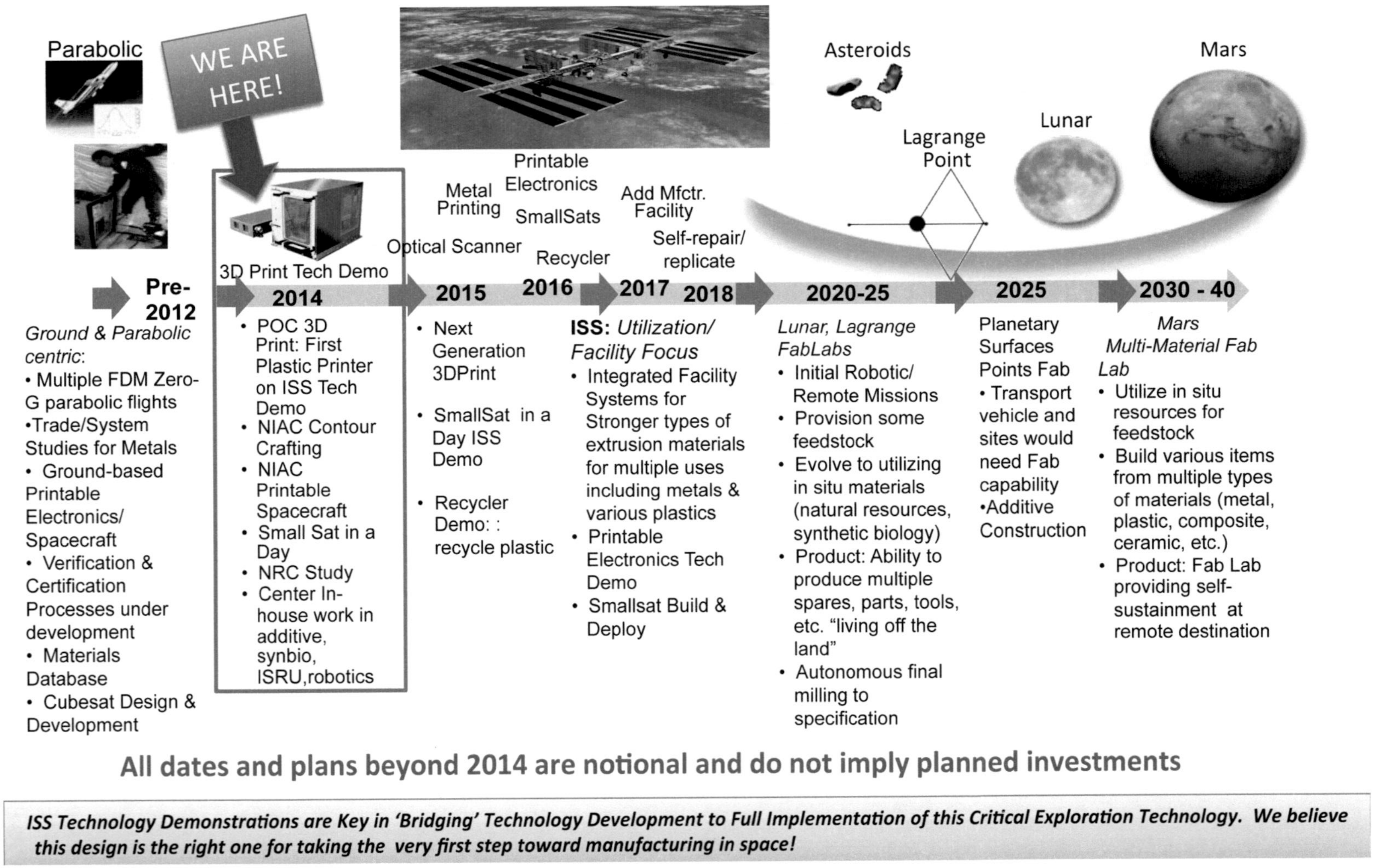


FIGURE 4.5 NASA Space Technology Mission Directorate's In-space Manufacturing Technology Development Vision, which can serve as a useful starting point for development of an agency-wide roadmap. SOURCE: NASA Marshall Space Flight Center.

# 5

# A Possible Way Ahead for the Air Force

Several Air Force commands have responsibilities for space systems, including the Air Force Space Command (AFSPC), headquartered at Peterson Air Force Base (AFB), and the Air Force Materiel Command, headquartered at Wright-Patterson AFB, both of which sponsored this study. Program management and system acquisition responsibilities for Air Force space systems are the responsibility of the Air Force Space and Missile Systems Center at Los Angeles AFB, which is a part of AFSPC. New developments in manufacturing technology for military space systems are the responsibility of the Materials and Manufacturing Directorate of the Air Force Research Laboratory (AFRL), located at Wright-Paterson AFB. AFRL's Space Vehicles Directorate, located at Kirkland AFB, engages in the discovery and delivery of technologies for satellite systems.[1]

Work done at AFRL includes initiatives in additive manufacturing, verification and validation related to manufacturing process simulation, and modeling related to aerospace systems.[2] This includes a history of additive manufacturing work with metals, including laser direct manufacturing; additive manufacturing of super alloys; and advanced manufacturing of specific alloys. Most of this work has been focused on aerospace aircraft applications, with only a modest amount of recent work directed toward using additive manufacturing for spacecraft construction or operations. The AFRL Space Vehicles Directorate engages in development of technologies related to spacecraft construction, systems, and operations.

Operational applications of additive manufacturing and associated technologies lie within the purview of the AFSPC and Air Force contractors working on specific launch and space systems, such as global positioning, wide-band global communications, missile warning, space and terrestrial weather, and strategic protected communications.

## THE CHALLENGE

The Air Force operates large and varied fleets of satellites. This responsibility now faces significant challenges in a period of constrained federal and Department of Defense (DOD) budgets. In 2011, the commander of AFSPC

[1] C.J. McNutt, R. Vick, H. Whiting, and J. Lyke, "Modular Nanosatellites-Plug-and-Play (PnP) Cubesat," AIAA 7th Responsive Space Conference 2009, Paper RS7-2009-4003, American Institute of Aeronautics and Astronautics (AIAA), Reston, Va., 2009.

[2] Air Force Research Laboratory (AFRL) documents often use "direct digital manufacturing" to mean "digital data to finished part with little human interaction," rather than the more widely used industry process term, additive manufacturing.

directed that priority be given to resolving four key challenges, one of which was to "provide a full-spectrum launch capability at dramatically lower cost."

This relates directly to the present study in that, if future additive manufacturing methods for constructing satellites on the ground and in space prove effective, the masses of satellites will likely be significantly lower than today, and the Air Force might be able to use smaller and less expensive launch vehicles to satisfy its mission needs. The high cost of launch vehicles is currently a major factor facing Space Command, and the Air Force has indicated that the reduction in launch costs is one of its primary incentives for looking into applications of additive manufacturing.[3]

With respect to valuable technologies for all areas of Air Force responsibility, the Office of the Air Force Chief Scientist issued a report in 2010 of a special study known as "Technology Horizons." This report provided Air Force and contractor organizations with a vision of science and technology (S&T) developments needed to support Air Force missions in the period 2010-2030.[4]

*Technology Horizons* made two pertinent observations with respect to Air Force space responsibilities. The first was that small satellites (i.e., those with mass less than 200 kg) might have useful military capabilities, including imaging, communication links, and scientific data about space weather.[5] The second observation was of the value of responsiveness to combatant commanders, aided perhaps by rapidly composable platforms put together with suitable optical, communications, or other systems and launched within several days in response to significant military threats.[6]

The topic of building spacecraft was raised 2 years later in the 2013 Air Force *Global Horizons* report.[7] This report provides the most recent, fully vetted public statement of expected Air Force challenges and opportunities over the coming three decades. It too was produced under the aegis of the Air Force's Office of the Chief Scientist, with active participation by numerous senior Air Force leaders, DOD military and civilian experts, and experts from other federal agencies and advisory committees.

*Global Horizons* spans many topics and includes evaluation of future technological capabilities available to the Air Force for fabricating aircraft and spacecraft using the tools of additive manufacturing. Additive manufacturing is now regarded as a potentially promising means for the Air Force and DOD to reduce the cost of designing and producing parts for air and space systems needed to fulfill Air Force missions. The report also considers the possibility of using additive manufacturing facilities for fabricating entire systems, such as drones and spacecraft. The possibility of on-orbit repair and maintenance using additive manufacturing technologies was also identified as a possible way to reduce the annual maintenance cost of defense satellite systems.

In attempting to address the issue of weight and cost growth, including mission growth, of the current fleet of Air Force satellites, the *Global Horizons* study reached the following conclusion:

> We can revolutionize our space architectures by using hosted payloads and launching smaller, affordable, and fractionated satellites in disaggregated constellations.[8]

This is a conclusion similar to that delineated in an AFSPC white paper, "Resiliency and Disaggregated Space Architectures."[9] Later, with respect to new technologies, the *Global Horizons* report concluded the following:

---

[3] Matt Fetlow, AFRL, presentation to the committee, April 17, 2014.

[4] Office of the U.S. Air Force Chief Scientist, *Technology Horizons (Final Report): Air Force Global Science and Technology Vision*, AF/ST TR 10-01-PR, September 2011; originally released on May 15, 2010, as *Report on Technology Horizons: A Vision for Air Force Science & Technology during 2010-2030,* Volume 1, AF/ST-TR-10-01-PR, Washington, D.C.

[5] Ibid, p. 33. The list of potential uses was considerably expanded in the Air Force's *Global Horizons* report, issued in 2013.

[6] Ibid, p. 69, see Potential Capability Area 27; p. 98.

[7] Office of the U.S. Air Force Chief Scientist, *Global Horizons; Air Force Global Science and Technology Vision* AF/ST TR 13-01, June 13, 2013.

[8] Ibid, p. 12.

[9] Air Force Space Command, "Resiliency and Disaggregated Space Architectures," white paper, released August 21, 2013, http://www.afspc.af.mil/shared/media/document/AFD-130821-034.pdf.

> New technologies such as additive manufacturing in space [enabling on-orbit construction and repair], combined with modular and open architectures, can help realize low-cost satellites, and agile, reconfigurable space systems.[10]

and suggested the Air Force

- Redefine space acquisition in accordance with disaggregated satellites and inexpensive launch with a goal of greater than 10X cost reduction employing advanced technologies,
- Pursue Air Force technologies for space, e.g., adaptive manufacturing in space, . . .
- Make targeted investments in autonomous/robotic systems and platforms.[11]

In 2013 the U.S. Air Force Scientific Advisory Board (AFSAB) undertook a study of the value of small satellites (having masses less than 300 kg) to the Air Force mission.[12] This topic was motivated by concern about a rapidly changing strategic setting for the Air Force's space systems. Such changes included evolving threats to U.S. military and intelligence spacecraft, diminishing federal budgets, increases of international space activity, increasing global capabilities for technology miniaturization, and emerging space launch options.

From their work, the AFSAB concluded that smaller satellites might be able to achieve significant defense mission capabilities within the next 2-5 years and serve important, continuing roles in the future. The AFSAB also recommended that the Air Force initiate various S&T investments to enable the far-term employment of smaller satellites. Details of these can be found in the abstract of their study.

There are several important technical opportunities for the Air Force that might contribute to the maintenance of space superiority in coming decades, as enumerated in both the *Technical Horizons* and *Global Horizons* reports and the work of the AFSAB. They are as follows:

- *Advanced manufacturing.* Advanced manufacturing technologies will enable open architectures that permit rapid prototyping, mission-specific reconfigurability; material tailoring for specific applications; efficient small-lot production; and better systems, faster and cheaper.
- *Redefined qualification and certification paradigm.* The qualification and certification paradigm can be redefined to allow rapid utilization of products from advanced manufacturing and additive manufacturing specifically (efficiently from prototype to practice). The new paradigm could eliminate the excessive development times for complex systems by inclusion of concepts such as defined and finite system life, qualification and certification as "adequate" for this application for this length of time, and process qualification and certification vis a vis component qualification and certification.
- *Digital Thread and Digital Twin.*[13] Digital Thread, comprised of advanced modeling and simulation tools that link materials, design, processing, and manufacturing information, will provide the agility and tailorability needed for rapid development and deployment, while also reducing risk. Digital Twin will be a virtual representation of the system as an integrated system of data, models, and analysis tools applied over the entire life cycle on a tail-number unique and operator-by-name basis. Modeling and simulation tools will optimize manufacturability, inspectability, and sustainability from the outset. Data captured from legacy and future systems will provide the basis for refined models that enable component and system-level prognostics.
- *Autonomous/remotely operated systems.* Home-station logistic operations and delivery will be enhanced with the increased use of robotic or remotely operated systems. Deploying the system should reduce the

[10] Ibid, p. 13.

[11] Ibid, p. 20.

[12] The AFSAB is a federal advisory committee that provides the Secretary of the Air Force and Chief of Staff of the Air Force with independent advice on Air Force science and technology. Such advice is based on annual studies of emerging technologies and their potential value to long-term needs of the Air Force. Air Force Scientific Advisory Board Study, "Microsatellite Mission Applications," July 2013. Note the definition of "microsatellite" differs from that of the Air Force Office of the Chief Scientist and Table 6.1, which is the nomenclature adopted by this committee.

[13] E.H. Glaessgen and D.S. Stargel, "The Digital Twin Paradigm for Future NASA and U.S. Air Force Vehicles," AIAA 53rd Structures, Structural Dynamics, and Materials Conference, April 26, 2012, AIAA, Reston, Va.

forward human footprint. Material processing and handling (armaments and cargo), servicing, maintenance, emergency response, protection, and base surveillance are all potential automation/remote operation targets.

- *On-site production.* Advances in manufacturing technology like additive manufacturing would allow rapid generation of needed devices and parts. Use of indigenous resources and assets, including recycled materials, offer flexible and potentially cost-saving procurement options.

## THE REALITY OF ADDITIVE MANUFACTURING

The committee understands the breadth and potential importance of broad forms of advanced manufacturing technology, and specifically additive manufacturing, to the Air Force's needs and responsibilities in space. Yet, additive manufacturing is still an emerging, incomplete, and relatively immature field of manufacturing technology, albeit rapidly evolving in commercial, academic, governmental, and entrepreneurial ground-based laboratories and facilities.

The cutting edge of additive manufacturing-related satellite system production is now at the level of creating and evaluating simple electromechanical systems for ground and space applications in small terrestrial laboratories across the United States and abroad.[14,15]

With respect to electronics, printing of electric power circuits with additive manufacturing is a viable undertaking, as evidenced by ongoing work over the past decade at the University of Texas, El Paso.[16] However, additive manufacturing is regarded as inadequate for printing of advanced digital electronic circuits that are essential for spacecraft. Commercial-quality additive manufacturing machines have minimum feature resolutions on the order of 50 to 100 μm. In contrast, radiation-hardened integrated circuits for space systems have feature sizes on the order of 0.35 μm.[17] Other very advanced circuits being considered for spacecraft design and implementation are at feature sizes of 90 nm and smaller. At the present time, the solution is to hand-insert and lock integrated circuits and other high-density digital circuit cards into additive manufacturing-prepared receptacles. In the future, a similar action will likely be the simplest solution for fabrication of spacecraft aboard a space platform, with the insertion being made by a local robot or some type of intelligent machine.

It is clear from these and other examples that many different questions of technology and engineering practice will have to be resolved before additive manufacturing can be embraced as an effective, dependable, cost-reducing, and strategically acceptable means of producing national security spacecraft on the ground, let alone in a human-tended or robotic orbiting facility.

## AIR FORCE EXPERIENCE WITH ADDITIVE MANUFACTURING

The Air Force relies on commercial vendors to design, build, test, and transport spacecraft into Earth orbit. Definitive requirements for the desired additive manufacture of spacecraft will have to be developed by the government before competitive contractual bidding can become a reality. Potential contractors will bid based on their understanding of all aspects of spacecraft documentation, construction, testing, evaluation, and operations as well as the contract incentives related to meeting cost, on-time delivery, lifetime on orbit, and other factors. At the present time, aerospace industry knowledge of additive manufacturing, while rapidly advancing, is in its early stages with respect to all aspects of spacecraft production, including physical and environmental properties, engineering and manufacturing, materials, knowledge and specification inserted electronics, and reliability.

The AFRL Materials and Manufacturing Directorate has a significant portfolio of work in additive manufacturing.[18] A lead AFRL researcher in additive manufacturing also serves as the Department of Defense manager for America Makes, a network of companies, nonprofit organizations, academic institutions, and government agencies that was founded in August 2012 as the flagship institute for other National Network for Manufacturing

[14] Aguilera et al., 3D Printing of Electro Mechanical Systems, SFF Symposium Proceedings, 2013.

[15] K. Short and D. Van Buren, Printable Spacecraft: Flexible Electronic Platforms for NASA Missions, Phase 1 Report, NIAC, September 2012.

[16] See http://engineering.utep.edu/announcement111813.htm.

[17] See http://www.atmel.com/Images/AERO-4015i-Integrated%20Circuits-Space%20Rad-Hard_US-E-0912_LR.pdf.

[18] Mary Kinsella, AFRL, presentation to the committee, April 17, 2014.

Innovation institutes. These institutes were established in response to the strategic guidance from the president to enhance U.S. capabilities and competitiveness in advanced manufacturing. While AFRL additive manufacturing work is extensive, it is, to date, largely aimed at aeronautical/aircraft applications (Figures 5.1 and 5.2). Significant additional research to fully close all the gaps will be required by AFRL to successfully implement additive manufacturing either in space or on the ground for space applications.

### Where Is Additive Manufacturing Most Beneficial?

Completely additively manufactured CubeSats are already being contemplated, and somewhat larger additively manufactured platforms could come as funds and capabilities expand. This work is important in that it provides the Air Force with valuable experience and data on additive manufacturing applications for space. However, as satellites tend toward larger and more complex (and operationally useful) designs, at some point, because of the differing size scales and accuracy requirements, the most efficient way to produce an entire spacecraft will involve multiple modes of construction. Some sections of the spacecraft will be most rapidly and effectively produced by additive manufacturing, while others will be most effectively made through extrusion, castings, or other manufacturing processes at another location. The fabrication of the complete satellite will involve assembly of these components at one site.

At the present time, it appears that additive manufacturing might be beneficial to all sizes of spacecraft, but the largest percentage mass and volume reductions will likely accrue to small satellites; that is, those with masses less than about 200 kg.

### The Air Force Vision: Fabrication of Spacecraft in Space

The Air Force charged the committee to, among other things, assess the feasibility of additively manufacturing a fully functional spacecraft and to identify S&T gaps needing to be filled to achieve such a goal. While there may be benefits to rapidly manufacturing an entire spacecraft at an orbital fabrication facility, it is not clear to the committee that such an achievement would be either operationally useful and desirable or economically feasible, especially in the short term (5-10 years).

> **Finding:** There is at present a lack of knowledge to credibly determine whether or not development of an Air Force-specific, space-based additive manufacturing production facility would achieve its expected benefit. Given that such a fabrication center would be highly complex and expensive, a detailed system assessment and cost-benefit analysis might be advisable.

Figure 5.3 depicts several of the areas of technology requiring additional research. When envisioning a capability to print an entire functional spacecraft using additive manufacturing technology in an orbiting facility, there are several questions that serve to illuminate this complexity. For example,

- Is the envisioned facility intended for a one-time use to be discarded after spacecraft production?
- If it is intended to be a multiuse facility capable of continually producing spacecraft over a period of years, will it produce a single satellite type or several satellite types?
- Will all of the spacecraft it produces be deployed in generally the same orbital inclination and orbital parameters? If not, what generic orbit would best suit the range of final orbit deployments?
- Would final deployment require an orbital transfer vehicle to provide relatively rapid deployment, or could a slower orbital-transfer subsystem be built into the final spacecraft (e.g., electric propulsion)?
- Is the size of the final spacecraft such that it could be produced internally within an additive manufacturing machine, or would it require an additive manufacturing machine that could produce satellites larger than itself?
- Would the selected technologies perform better in an atmosphere or in a vacuum?
- As an important, space-based site of Air Force resources, can this facility be adequately defended against destructive actions on the part of other nations or other aggressors?

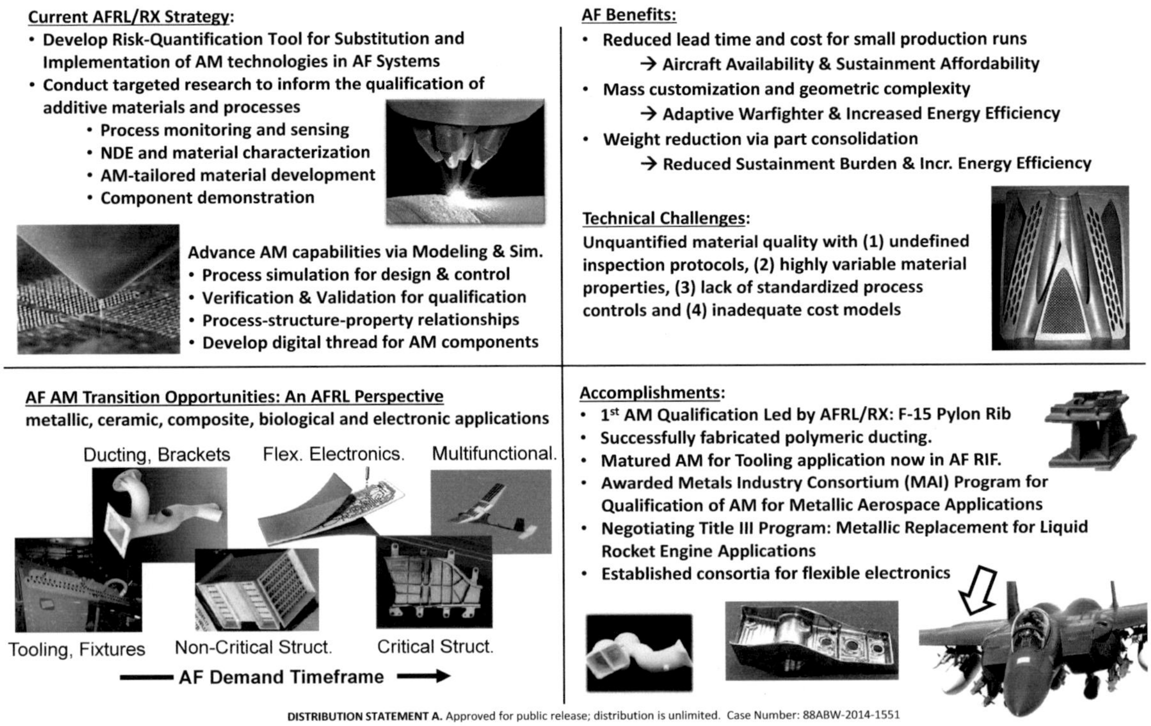


FIGURE 5.1 The Air Force Research Laboratory's Additive Manufacturing Strategy emphasizes the development of this technology primarily for ground-based use for aircraft. SOURCE: Courtesy of the U.S. Air Force.

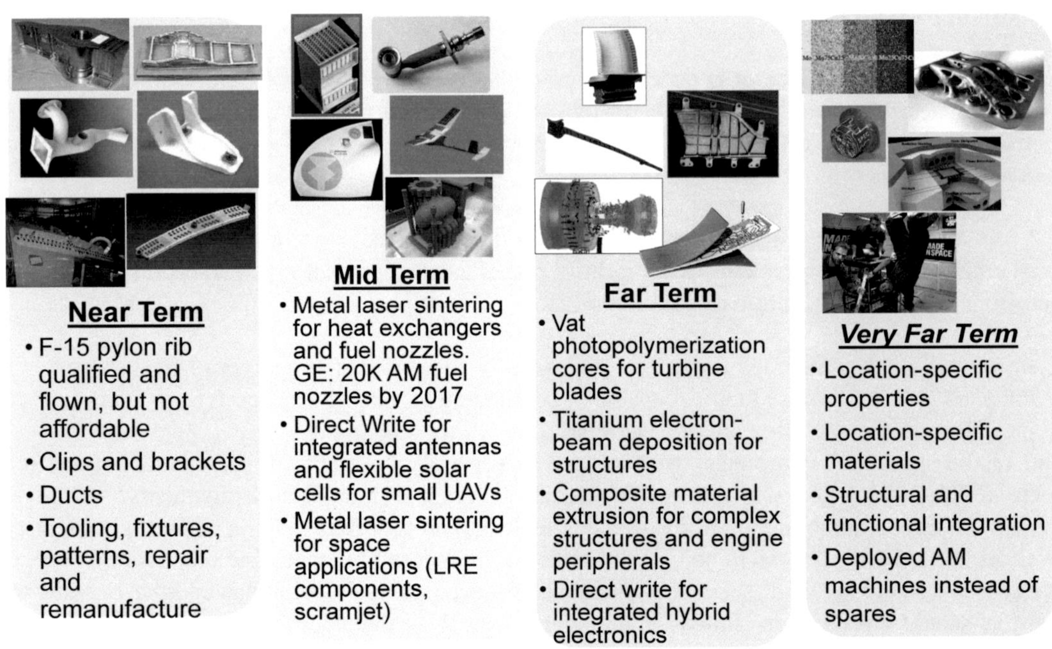


FIGURE 5.2 Currently, the Air Force Research Laboratory (AFRL) is focusing on ground-based and aviation-related additive manufacturing technologies and applications. AFRL has evaluated relatively few space-related applications. SOURCE: Courtesy of the U.S. Air Force.

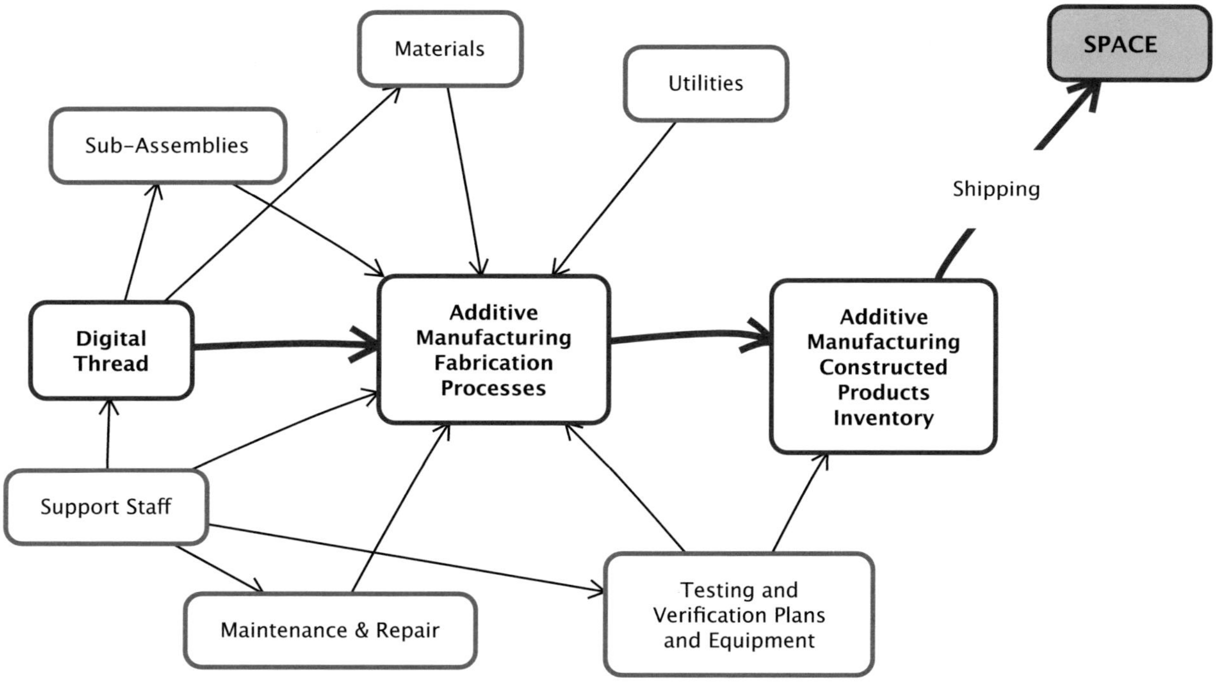


FIGURE 5.3 Areas requiring further research for development of a space-based additive manufacturing production facility.

The answers to these questions would directly influence the type and size of the required additive-manufacturing machine and supporting infrastructure.

Producing an entire functional spacecraft would require a number of different materials to produce the desired component properties and capabilities. This could require multiple writing heads, which would need to be precisely controlled in three dimensions over the entire production volume. There are certain components whose characteristics make producing them with additive manufacturing technologies extremely difficult, if not impractical. Two examples are microelectronics and optics. It is highly unlikely that any reasonable amount of technology investments in additive manufacturing would ever result in a competitive alternative to the current microelectronics lithography capabilities. Similarly, the precise accuracies required in optics would be difficult to replicate with additive manufacturing processes. As an example, a star sensor is a basic component of most satellites and is relatively unsophisticated compared to many other satellite subsystems. Yet, it would be extremely difficult to manufacture all of the components of a precision star sensor using additive manufacturing on the ground, let alone in orbit. Additive manufacturing will not replace microelectronics fabrication, but the committee believes that it does not have to. Additive manufacturing can print the structure and the electronic and other components required for functionality, manufactured by other means, can be embedded as described in earlier sections of this report.

Once the additive manufacturing system's details have been worked out, the support services and utilities required to establish a functional production facility would need to be addressed. Again important questions would need to be resolved. For instance,

- What structure would be required to support the machine and all its required support functions?
- How much power would the facility require in both its operating and dormant modes?
- How would raw materials be attached, stored, repositioned, mounted onto the additive manufacturing machine, and removed after use?

- How would the resupplied raw materials be brought into proximity to the production facility and attached?
- What type of attitude control and vibration control would the facility require to support the additive manufacturing machine?
- How do the construction machines test, detect, and repair errors in the fabricated items they produce?
- Would the facility need a propulsion capability to maintain its orbit?
- Would the production facility be able to function entirely autonomously, or would it require continual human tending? If so, would this human tending require an associated habitat to support the crew? Could it be human tended only during production runs? Could it be run autonomously and only require periodic human maintenance?

Finally, with all these technical and operational questions answered, the Air Force might be in a position to determine if the system would provide sufficient economic advantages to merit deployment and, more importantly, if it would provide sufficient new military capabilities or operational advantages to merit its cost. If the answers to all of these questions were judged to be affirmative, it would still need to be determined if the system would bring with it military vulnerabilities that could be effectively be overcome. Such a production facility would be vulnerable to both natural and adversarial hazards as it continually orbited Earth.

**Recommendation: As the technology evolves and as projects utilizing this technology are considered, the Air Force should conduct a systems-analytical study of the operational utility of spacecraft and spacecraft components produced in space using additive manufacturing compared to other existing production methods.**

## A WAY AHEAD FOR THE AIR FORCE

Additive manufacturing is a technology that has great potential to significantly reduce payload mass and size of national security spacecraft and, thereby, achieve a lower unit cost per spacecraft. However, this technology is in its infancy and, as yet, has not achieved sufficient technological maturity to be an immediate alternative to traditional fabrication of national security spacecraft.

The discussion presented here provides a path for the Air Force to begin to answer some of the fundamental questions of economic and operational benefit.

### Some In-Space Specific Technical Challenges

Should the Air Force ultimately decide to pursue additive manufacturing in space, there are unique technical challenges that will have to be overcome. In this context, the assertion that bringing raw materials to space could significantly reduce overall costs of putting a new spacecraft into use is only a small part of the overall operating expenses of such a facility.

Additive manufacturing by itself is a relatively slow, energy-intensive construction process. Assembling a spacecraft on the ground would likely be faster, and the fastest approach may well be rapid construction in a ground facility, followed by a rapid launch from a "smaller" launch vehicle requiring a minimum of integration effort. Determining how to avoid long satellite construction time is essential before committing to construction of an in-space additive manufacturing system. (This argument is also true for an additive manufacturing repair platform.)

The designs and equipment fabrication for Air Force missions are done with careful attention to materials selection and mechanical, electrical, and thermal standards to enable the equipment to survive the strong forces and rapidly changing thermal environment of the launch vehicle. Once in space, the diverse array of space equipment will be exposed to a complex mix of environmental conditions, including the near-vacuum of space, the strong, orbitally varying heating effects of sunlight, possible electrostatic charging, deleterious effects of strong solar ultraviolet radiation on the equipment container's thermal protection and surface materials, reduction of efficiency of solar electric cells resulting from solar radiation and highly variable energetic particle fluxes, and the damage that occasionally arises from micro-meteorites.

Equipment sent into space as a free-flying platform is designed as a self-consistent, operational entity with many different systems and subsystems cooperatively working in synchronism to achieve the goals of its architects. There is a vast range of specialized materials and topological complexity in such objects, including both the components of the satellite system operations and the highly complex and often extraordinarily miniaturized components of scientific instruments. In addition, scientific instruments sent into space have to be carefully calibrated and operationally tested in special ways prior to launch, often precluding any easy way to recalibrate such instruments while in space.

This is not to say that a major investment in space-based additive manufacturing should not be made. Rather, it is clear that additive manufacturing offers niche advantages to space systems, both for ground-based engineering and space-based operations. But a long-term, strategic plan of engineering system and operation planning investments is essential to take advantage of this new manufacturing technology in the environment of space. These should include many different efforts to build space-qualified parts and subsystems via additive manufacturing prototyping and even final flight-worthy components.

**Finding:** An independent, free-flying additive manufacturing satellite construction platform, human-tended or robotic, will require extensive ground-based development in additive manufacturing, robotics, and telepresence. Given the various limitations of power, cost, long build times, verification of manufacture, and other factors discussed previously, a large number of issues require resolution before committing to such a program.

## ADDITIVE MANUFACTURING FOR SPACE

It appears to the committee that the ability to develop a space-based additive manufacturing capability able to produce fully functional, operational Air Force satellites in orbit anytime in the reasonable future is well beyond the current state of the art or, for that matter, any current technology plans. There are, however, several key technology areas that are unlikely to be pursued by either commercial industry or NASA that the Air Force could reasonably fund that would contribute to a better understanding of the eventual feasibility of such a capability. Some examples of these are discussed in Chapter 3. They include robotics in zero gravity and materials processing in zero gravity and in a vacuum. Additionally, there are a number of activities that should be pursued to qualify additive manufactured materials for application in the space environment. The detailed characterization of all these materials and their properties over time in the environments they would be exposed to in space need to be well understood. This is important not only for their considered use in operational systems, but also to understand what, if any, potential vulnerabilities they might introduce into operational systems.

At the moment, the economic drivers for investigating some of the issues discussed above (the in situ manufacturing of small, complex electronic and optical parts, as well as motors) are minimal; there is not strong "push" in these areas to include these types of specialized components in additive manufacturing studies.

**Recommendation: The Air Force should continue to invest in additive manufacturing technologies, with a specific focus on their applicability to existing and new space applications, and invest in selected flight experiments.**

The committee believes it to be very important for the Air Force technology communities to stay acutely aware of all the activity and progress in the additive manufacturing area and its potential for space applications; without this level of currency, the Air Force will not be able to be an effective and knowledgeable consumer of this potentially important capability.

**Recommendation: The Air Force should consider additional investments in the education and training of both materials scientists with specific expertise in additive manufacturing and spacecraft designers and engineers with deep knowledge of the use and development of additively manufactured systems.**

Further, in the near term, the Air Force should not lose sight of the opportunities to apply additive manufacturing capabilities to existing or pending space systems. The introduction of materials and components produced with additive manufacturing techniques could reduce the cost, lead-time, weight, or other important factors.

As described above, the challenges of manufacturing of entire spacecraft in space are daunting, and the benefits are unclear. The technology challenges enumerated in Chapter 3 would form the basis for a long-term technology investment plan. However, the benefit of additive manufacturing in ground-based production is a far more demonstrated fact. In addition, there may be cases in the near- or midterm future where the space-based manufacture of less complex components and subsystems for spacecraft, or space-based assembly, may have operational and economic benefit (for example, in the construction of large antenna apertures, which have related challenges such as maintaining precise attitude pointing, establishing surface precision, and ensuring structural stability). If the Air Force can consider and take advantage of opportunities to incorporate additive manufacturing, then technical, acquisition, and contracting policies can provide incentives to Air Force contractors for performing the necessary research and to incorporate additively manufactured parts in space and launch systems. The committee cannot perform this task for the Air Force because the military has to develop its own requirements and bring together the interested parties that will implement a research strategy.

**Recommendation: The Air Force should establish a roadmap with short- and longer-term goals for evaluating the possible advantages of additive manufacturing in space. The Air Force should build on the considerable experience gained from other Air Force technology development roadmaps. The space-based additive manufacturing roadmap should include, but not be limited to the following:**

- **Developing goals for using the technology in key Air Force missions, especially for autonomously or minimally attended, space-based additive manufacturing and free-flyer missions;**
- **Identifying flight opportunities, including those on non-Air Force platforms such as on the International Space Station during its next decade of operations;**
- **Targeting the full technology-development life-cycle and insertion strategies through 2050, aligned with Air Force missions, and related collaborations.**

NASA is currently the leader in the development of space-based additive manufacturing technology and also operates a valuable research platform, the International Space Station (ISS). The ISS offers both internal and external research locations. For example, it may be possible to mount an additive manufacturing machine, such as a machine for producing large truss or antenna structures, to an external location on the ISS and operate it in the vacuum environment while enabling close monitoring and later inspection of parts. DOD has a history of ISS research, typically participating through the Air Force's Space Test Program. The Air Force can maximize its return on investment by seeking cooperative opportunities with NASA whenever possible.

**Recommendation: The Air Force should make every effort to cooperate with NASA on in-space additive manufacturing technology development, including conducting research on the International Space Station.**

**Recommendation: NASA and the Air Force should jointly cooperate, possibly involving additional parties including other government agencies as well as industry, to research, identify, develop, and gain consensus on standard qualification and certification methodologies for different applications. This cooperation can be undertaken within the framework of a public/private partnership such as America Makes.**

Cooperation can take many forms. The Air Force and NASA could jointly share the costs of research and development, or merely share data. But the opportunities available are greater now than they have been in even the recent past.

## CONCLUSION

Additive manufacturing technology is developing rapidly, so rapidly that it is difficult to determine its applications just a few years in the future. However, the committee concluded that the Air Force has been paying less attention to this developing technology for space use than it should, and it may be missing opportunities to leverage the work that is being conducted by NASA. By starting efforts to consider where additive manufacturing technology can possibly fit into its existing missions, and where it might have positive benefits for things such as reducing launch costs, the Air Force may identify unique value and encourage those actively involved in this technology development to propose new solutions to the Air Force's space requirements.

# Appendixes

# A

# Committee Biographical Information

ROBERT H. LATIFF, *Chair*, is president and a consultant of Latiff and Associates. Previously, General Latiff has served as vice president, chief engineer, and technology officer in SAIC's space and geospatial intelligence business unit. He is retired from the U.S. Air Force as a Major General, with his last assignments at the National Reconnaissance Office as the director for systems engineering and as the director of advanced systems and technology. General Latiff was a career acquisition officer, managing large complex systems such as the Cheyenne Mountain Complex, the Air Force's airspace management and landing systems, and the Joint Strategic Target Attack Radar System. Dr. Latiff holds M.S. and Ph.D. degrees in materials science and a B.S. in physics from the University of Notre Dame. He has previously served on the National Research Council's (NRC's) National Materials and Manufacturing Board (currently as a member and previously as chair), as a member of the Air Force Studies Board, and as chair of the Committee on Defense Materials, Manufacturing and Infrastructure. He has also served as chair of the Committee on Materials and Manufacturing Sustainability for Department of Defense Systems: A Workshop and the Committee on Assessing the Need for a Defense Stockpile.

PETER M. BANKS is a partner at Red Planet Capital Partners. Prior to the creation of Red Planet Capital Partners, Dr. Banks was CEO and president of ERIM International, Inc., and partner at XR Ventures, the investment arm of the X-Rite Corporation. Previously, he served as dean of engineering at the University of Michigan and was a faculty member of Stanford University and the University of California. Dr. Banks earned his Ph.D. in physics at Pennsylvania State University following an M.S.E.E. degree from Stanford University. While at Stanford, he led the Space, Telecommunications and Radio Science Laboratory of the Department of Electrical Engineering. He participated in several space shuttle missions as a principal investigator for the Shuttle Electro-Dynamic Tether System and as a co-investigator on the flight of Spacelab-1. He has received the U.S. government's Distinguished Public Service Medal for his contributions to NASA programs. Dr. Banks is a member of the National Academy of Engineering (NAE) and served as co-chairman of the NRC's former Commission on Physical Sciences, Mathematics, and Applications. He served as the chair of the board of the Universities Space Research Association and on the boards of a number of start-up companies, including Triformix (Santa Rosa, California) and HandyLab (Ann Arbor, Michigan). Dr. Banks has advised the Euro-America funds and various federal agencies for work related to defense, space exploration, and national economic security. He has served on the NRC's Report Review Committee and the Committee on Assessment of NASA Laboratory Capabilities.

ANDREW S. BICOS is director of enterprise manufacturing technology in the Office of the CTO at Boeing Company. Dr. Bicos is responsible for developing the enterprise strategy for all manufacturing research and development (R&D) at Boeing and managing the portfolio of activities that develop and transition technologies and processes into Boeing's factories and wide array of products, focusing on research into composites, automated and robotic fabrication and assembly, additive manufacturing, and factory ergonomics and worker safety. Prior to this assignment, he was the director for enterprise structures technology, and before that he was director of Phantom Works' materials, structures, and manufacturing technology thrust. Dr. Bicos has published more than 20 technical papers and articles on innovative composites, adaptive structures, and vibration reduction technologies. He has held positions on the American Institute of Aeronautics and Astronautics (AIAA) Structural Dynamics and American Society of Mechanical Engineers (ASME) Adaptive Structures and Material Systems technical committees and is a former chair of the ASME Aerospace Division. He is an associate fellow of the AIAA. Dr. Bicos is currently the chair of the ASME Aerospace R&T Task Force and is the Boeing executive focal for ASME and sits on the ASME Industry Advisory Board where he is the chair of the executive committee. Dr. Bicos is also the Boeing executive focal for California State University, Los Angeles. He has a Ph.D. in aeronautical engineering from Stanford and an M.B.A. from University of California, Los Angeles.

ELIZABETH R. CANTWELL is the director for economic development at Lawrence Livermore National Laboratory. She previously served as the deputy associate laboratory director for the National Security Directorate at Oak Ridge National Laboratory. Prior to joining Oak Ridge, Dr. Cantwell was the division leader for the International, Space, and Response Division at Los Alamos National Laboratory. Her career began in building life-support systems for human spaceflight missions with NASA. She received an M.S. in mechanical engineering from the University of California, Berkeley, an M.B.A. in finance from the Wharton School of the University of Pennsylvania, and a Ph.D. in mechanical engineering from the University of California, Berkeley. Dr. Cantwell has extensive NRC experience, including current memberships on the Space Studies Board and the Division on Engineering and Physical Sciences Committee; co-chair of the Committee on Decadal Survey on Biological and Physical Sciences in Space; and member of the Committee on NASA's Bioastronautics Critical Path Roadmap, the Review of NASA Strategic Roadmaps: Space Station Panel, the Committee on Technology for Human/Robotic Exploration and Development of Space, and the Committee on Advanced Technology for Human Support in Space.

RAVI B. DEO is president of EMBR, a small business specializing in the design and manufacturing technology development of composite structures. He has worked as a program and functional manager for government sponsored projects on cryotanks, integrated system health management, aerospace structures, materials, subsystems, avionics, thermal protection systems, and software development. He has extensive experience in road-mapping technologies, program planning, technical program execution, scheduling, budgeting, proposal preparation, and business management of technology development contracts. Among his significant accomplishments are the NASA-funded SLI, NGLT, OSP, and High Speed Research programs where he was responsible for the development of multidisciplinary technologies. Dr. Deo is the author of more than 50 technical publications and is the editor of one book. He has served on the NRC steering committee for NASA Space Technology Roadmaps and Priorities and on Panel C: Structures and Materials of the Steering Committee on Decadal Survey of Civil Aeronautics and Panel J: High-Energy Power and Propulsion and In-space Transportation of the Committee for the Review of NASA's Capability Roadmaps. He has also served on the Scientific Advisory Board to the Air Force Research Laboratories. He was formerly director of the Technology, Space Systems Market Segment at Northrop Grumman Corporation's Integrated Systems Sector through 2008. Dr. Deo received his bachelor's degree in aeronautical engineering from Indian Institute of Technology in Bombay, India, and his master's and Ph.D. degrees in aerospace engineering from Georgia Institute of Technology. He is a current member of the Aeronautics and Space Engineering Board.

JOHN W. HINES is an independent consultant and senior technology advisor specializing in the areas of space technologies, medical and biological technologies, technology aggregation, and technology program/project management. He recently retired as the chief technical officer for the NASA Ames Research Center (ARC). In that capacity, he identified, defined, developed, and integrated transformational space technologies for NASA and national goals and objectives through the ARC Office of the Center Director and the NASA chief technologist. Prior to this, Mr. Hines was chief technologist in the ARC Engineering Directorate, and before that he was deputy chief and chief technologist for the Small Spacecraft Division. He has more than 40 years of combined NASA, Air Force, and research center experience in biological and biomedical technology development, satellite/spaceflight hardware development, electronic systems engineering, program/project management, advanced technology assessment and development, and technology/program advocacy. He has a B.S. in electrical engineering from Tuskegee University and a M.S. in biomedical and electrical engineering from Stanford University.

BHAVYA LAL is a research staff member at the IDA Science and Technology Policy Institute (STPI) where her research and analysis focuses on space technology and policy and is frequently incorporated in national policy documents. Recent and ongoing projects include supporting the Office of Science and Technology Policy in developing a national space technology strategy, improving detection of near Earth objects, documenting global trends in space, and examining recent commercial activities in space including their legal ramifications related to the Outer Space Treaty. Before joining STPI, Dr. Lal was president of C-STPS, LLC, a science and technology policy research and consulting firm in Waltham, Massachusetts. Prior to that, she was the director of the Center for Science and Technology Policy Studies at Abt Associates, Inc., in Cambridge, Massachusetts. Dr. Lal holds B.S. and M.S. degrees in nuclear engineering from the Massachusetts Institute of Technology (MIT), an M.S. from MIT's Technology and Policy Program, and a Ph.D. from the Trachtenberg School of Public Policy and Public Administration (concentration in science and technology policy) at George Washington University.

SANDRA H. MAGNUS is the executive director of the AIAA. Selected to the NASA Astronaut Corps in April 1996, Dr. Magnus flew in space on the STS-112 space shuttle mission in 2002 and on the final space shuttle flight, STS-135, in 2011. In addition, she flew to the International Space Station (ISS) on STS-126 in November 2008, served as flight engineer and science officer on Expedition 18, and returned home on STS-119 after 4½ months on board. Following her assignment on the ISS, Dr. Magnus served at NASA Headquarters in the Exploration Systems Mission Directorate. Her last duty at NASA, after STS-135, was as the deputy chief of the Astronaut Office. While at NASA, Dr. Magnus worked extensively with the international community, including the European Space Agency and the National Space Development Agency of Japan, as well as with Brazil on facility-type payloads. She also spent time in Russia developing and integrating operational products and procedures for the ISS. Before joining NASA, Dr. Magnus worked for McDonnell Douglas Aircraft Company as a stealth engineer. While at McDonnell Douglas, she worked on internal R&D and on the Navy's A-12 Attack Aircraft program studying the effectiveness of radar signature reduction techniques. Dr. Magnus has received numerous awards, including the NASA Space Flight Medal, the NASA Distinguished Service Medal, the NASA Exceptional Service Medal, and the "40 at 40 Award" (given to former collegiate women athletes to recognize the impact of Title IX). Dr. Magnus has an M.S. in electrical engineering from Missouri University of Science and Technology and a Ph.D. from the Georgia Institute of Technology for materials science and engineering.

THOMAS E. MAULTSBY is the founder and president of Rubicon, LLC, a referral-based aerospace and technology consulting company. Mr. Maultsby has more than 44 years of space and technology experience in a variety of government and industry positions. His current focus is on space studies and analyses and independent program reviews. Past activities include space shuttle operations at NASA Headquarters, expendable launch operations at Vandenberg Air Force Base, satellite production and test operations, development and acquisition of nuclear treaty monitoring systems, and advanced technology development and insertion. His launch vehicle experience includes the space shuttle, Titan, Delta, Atlas, and Sea Launch. He has led or participated in several senior-level program reviews, including space shuttle operations for the NASA administrator, the Cassini program for the director of the Jet Propulsion Laboratory, the Defense Science Board Task Force on Space Superiority, and the NRC review of the NASA Communications Program. Mr. Maultsby has been a board member and chairman of the board of the Security Affairs Support Association (now the Intelligence and National Security Alliance), a member of the AIAA Space Transportation Technical Committee, and a member of the AeroAstro board of directors. He is a co-founder of the Small Payload Ride Share Association and organizes the programs for the annual conferences.

Mr. Maultsby is an associate fellow of the AIAA. He has an M.S. in engineering from the University of Louisville and an M.S. in systems management from the University of Southern California. He has served on the NRC's Committee to Review NASA's Space Communication Program.

MICHAEL T. McGRATH is the engineering director at the Laboratory for Atmospheric and Space Physics (LASP) and a professor adjunct in Aerospace Engineering Sciences at the University of Colorado, Boulder. His experience in mechanical design for space began at LASP with the Ultraviolet Spectrometer for Pioneer Venus and the Photopolarimeter for Voyager. He was technical program manager for the Solar Mesospheric Explorer instrument module and spectrometers. He then joined the High Altitude Observatory at the National Center for Atmospheric Research with roles in group and section management that included instrument development for the Mauna Loa Solar Observatory, the development of the SPARTAN 201 White Light Coronagraph, and management of the successful repair-in-space of the Solar Maximum Mission Coronagraph/Polarimeter. While there he supported two successful expeditions to photograph total eclipses in Indonesia and the Philippines. He was the U.S. project manager supporting the successful collaborative proposal for the US/UK HIRDLS project. He then returned to LASP as mechanical engineering manager to oversee the development of the CASSINI UVIS instrument currently at Saturn and the TIMED pointing platform. He architected the design of the Student Nitric Oxide Explorer spacecraft, supported the Spectral Intensity Monitor and Total Intensity Monitor instrument developments for the Solar and Climate Experiment (SORCE) mission, and then as engineering director, along with responsibility for all projects in LASP engineering, was project manager for NASA's Aeronomy of Ice in the Mesosphere Small Explorer mission that is currently in extended mission. Mr. McGrath has 40 years of experience in the development of sensor and instrument systems for NASA's sounding rocket, Earth, and planetary programs. He has received 13 NASA Group Achievement Awards and served on NASA's Technology Management Working Group and NASA's Science Definition Team for Student Collaborations. As professor adjunct he teaches Introduction to Engineering Projects in the school of engineering's Integrated Teaching and Learning Laboratory—recipient of the Gordon Prize for innovative curriculum development. For two decades he has taught a graduate class in aerospace engineering in spacecraft design. He has a B.S. in engineering from the University of Colorado, Boulder.

LYLE H. SCHWARTZ is a retired director of the Air Force Office of Scientific Research where he had responsibility for the basic research program of the Air Force. He began his career as a professor of materials science and engineering at Northwestern University where he also became director of the Materials Research Center. He later became director of the Materials Science and Engineering Laboratory at the National Institute of Standards and Technology where he was responsible for management of the R&D agenda for metals, ceramics, polymers, magnetic materials, and development and standardization of techniques for materials characterization. Dr. Schwartz subsequently assumed responsibility for basic research on structural materials of interest to the U.S. Air Force, in addition to the areas of propulsion, aeromechanics, and aerodynamics. His current interests include government policy for R&D and science, technology, engineering, and mathematics education in grades 6-12 via materials science/technology. Dr. Schwartz received both his B.S. in engineering and his Ph.D. in materials science from Northwestern University. He is a member of the NAE and has served on numerous NRC committees and boards, including as vice chair of the National Materials Advisory Board, as chair of the Army Research Laboratory Technical Advisory Board, and as a member of the Air Force Studies Board. Recent NRC committee membership included the following studies: Manufacturing Program at the National Institute of Standards and Technology (NIST), Best Practices in Assessment of Research and Development Organizations, and Examination of the U.S. Air Force's Aircraft Sustainment Needs in the Future and Its Strategy to Meet Those Needs.

IVAN E. SUTHERLAND is a visiting scientist at Portland State University and where he works in the Asynchronous Research Center that he founded with Marly Roncken in 2008. His research has focused on the design of self-timed or asynchronous computer circuits and computer architecture, and his interests include pursuing creative contributions in computer science and computer graphics. Dr. Sutherland is a member of both the National Academy of Sciences and the NAE. He is the 1988 recipient of the Turing award and the 2012 recipient of the Kyoto Prize in Advanced Technology. He is author of more than 60 patents, as well as numerous papers. Dr. Sutherland received a Ph.D. from MIT in electrical engineering, and he holds honorary degrees from Harvard University, the University of North Carolina, the University of Utah, and Carnegie Mellon University. He has served on the NRC's Committee on High Performance Computing and Communications: Status of a Major Initiative and NRC's Selection Committee for the Greatest Engineering Achievements of the 20th Century.

RYAN WICKER is a professor of mechanical engineering and director and founder of the W.M. Keck Center for 3D Innovation at the University of Texas, El Paso, where he also holds the endowed Mr. and Mrs. MacIntosh Murchison Chair I in Engineering. The Keck Center represents a world-class research facility that focuses on the use and development of additive manufacturing technologies for fabricating 3D objects that are plastic, metal, ceramic, bio-compatible materials, composite materials, or that contain electronics. Major research efforts are underway at the Keck Center in the areas of additive manufacturing technology development; closed-loop process control strategies for additive manufacturing; additive manufacturing of various powder metal alloy systems; and 3D structural electronics in which electronics, and thus intelligence, are fabricated within additive manufacturing-fabricated mechanical structures. Dr. Wicker received degrees in mechanical engineering from the University of Texas, Austin (B.S.) and Stanford University (M.S., Ph.D.).

PAUL K. WRIGHT is the director for Center for Information Technology in the Interest of Society (CITRIS) at University of California, Berkeley. CITRIS serves four University of California campuses and hosts many multidisciplinary projects on large societal problems, including healthcare, services, and intelligent infrastructures such as energy, water, and sustainability. Dr. Wright teaches in the Mechanical Engineering Department, where he holds the A. Martin Berlin Chair. He also serves as co-director of the Berkeley Manufacturing Institute and co-director of the Berkeley Wireless Research Center. From 1995 to 2005 he served as co-chair of the Management of Technology Program (a joint program with the Haas School of Business). His research and teaching are in high-tech product design and rapid manufacturing. Currently, he and his colleagues are designing and prototyping wireless systems for "Demand Response Power Management" throughout California, funded by the Public Interest Energy Research program of the California Energy Commission. Previously he served in faculty positions at New York University and Carnegie Mellon University. He received his B.S. and Ph.D. in industrial metallurgy from the University of Birmingham. Dr. Wright is a member of the NAE, and he served on the NRC's Committee on 21st Century Manufacturing: The Role of the Manufacturing Extension Partnership Program of the National Institute of Standards and Technology and the Panel on Manufacturing Engineering—2010.

### *Staff*

DWAYNE A. DAY, *Study Director*, a senior program officer for the NRC's Aeronautics and Space Engineering Board (ASEB), has a Ph.D. in political science from the George Washington University. Dr. Day joined the NRC as a program officer for the Space Studies Board (SSB). Before this, he served as an investigator for the Columbia Accident Investigation Board, was on the staff of the Congressional Budget Office, and also worked for the Space Policy Institute at the George Washington University. He has held Guggenheim and Verville fellowships and was an associate editor of the German spaceflight magazine *Raumfahrt Concrete,* in addition to writing for such publications as *Novosti Kosmonavtiki* (Russia), *Spaceflight,* and *Space Chronicle* (United Kingdom). He has served as study director for several NRC reports, including *Space Radiation Hazards and the Vision for Space Exploration* (2006), *Grading NASA's Solar System Exploration Program: A Midterm Review* (2008), and *Opening New Frontiers in Space: Choices for the Next New Frontiers Announcement of Opportunity* (2008).

ERIK SVEDBERG is currently a senior program officer of the National Materials and Manufacturing Board at the National Academies. In this role, he works with experts from across the nation to develop, negotiate, and oversee scientific and technical advisory studies for federal agencies to address questions about materials science, manufacturing, and engineering design. His activities at the National Academies have included work as a study director for studies on research opportunities in science and engineering, materials needs and R&D strategy for future propulsion systems, grand challenges in corrosion research, opportunities in protection materials, triennial review of the National Nanotechnology Initiative, and optics and photonics. Dr. Svedberg has a decade of industry experience with both small and large companies in the materials science area and was also a guest researcher at NIST for several years. He has been awarded and overseen many research grants and has published more than 80 scientific articles, been granted two patents, and his work is cited more 500 times. He holds both a master's and Ph.D. degree from the Department of Physics, Chemistry and Biology at Linköping University in Sweden and was the year 2000 recipient of the International Union for Vacuum Science, Technique and Applications Welch Scholarship.

ANDREA M. REBHOLZ, program coordinator, joined the ASEB in 2009. She began her career at the National Academies in October 2005 as a senior program assistant for the Institute of Medicine's Forum on Drug Discovery, Development, and Translation. Prior to the Academies, she worked in the communications department of a D.C.-based think tank. Ms. Rebholz graduated from George Mason University's New Century College in 2003 with a B.A. in integrative studies-event management. She earned the Certified Meeting Professional designation in 2013 and has more than 11 years of experience in event planning.

MICHAEL H. MOLONEY is the director for Space and Aeronautics at the SSB and the ASEB of the National Research Council of the U.S. National Academies. Since joining the ASEB/SSB, Dr. Moloney has overseen the production of more than 40 reports, including four decadal surveys—in astronomy and astrophysics, planetary science, life and microgravity science, and solar and space physics—a review of the goals and direction of the U.S. human exploration program, a prioritization of NASA space technology roadmaps, as well as reports on issues such as NASA's Strategic Direction, orbital debris, the future of NASA's astronaut corps, and NASA's flight research program. Before joining the SSB and ASEB in 2010, Dr. Moloney was associate director of the BPA and study director for the decadal survey for astronomy and astrophysics (Astro2010). Since joining the NRC in 2001, Dr. Moloney has served as a study director at the National Materials Advisory Board, the Board on Physics and Astronomy (BPA), the Board on Manufacturing and Engineering Design, and the Center for Economic, Governance, and International Studies. Dr. Moloney has served as study director or senior staff for a series of reports on subject matters as varied as quantum physics, nanotechnology, cosmology, the operation of the nation's helium reserve, new anti-counterfeiting technologies for currency, corrosion science, and nuclear fusion. In addition to his professional experience at the National Academies, Dr. Moloney has more than 7 years' experience as a foreign-service officer for the Irish government—including serving at the Irish Embassy in Washington and the Irish Mission to the United Nations in New York. A physicist, Dr. Moloney did his Ph.D. work at Trinity College Dublin in Ireland. He received his undergraduate degree in experimental physics at University College Dublin, where he was awarded the Nevin Medal for Physics.

# B

# Acronyms

| | |
|---|---|
| 3D | three-dimensional |
| AFB | Air Force Base |
| AFRL | Air Force Research Laboratory |
| AFSAB | Air Force Scientific Advisory Board |
| AFSPC | Air Force Space Command |
| ASI | Italian Space Agency |
| CAD | computer-aided design |
| DARPA | Defense Advanced Research Projects Agency |
| DOD | Department of Defense |
| ESA | European Space Agency |
| EU | European Union |
| FDM | fused-deposition modeling |
| GSFC | Goddard Space Flight Center |
| HEOMD | Human Exploration and Operations Mission Directorate |
| ISO | International Organization Standardization |
| ISS | International Space Station |
| JPL | Jet Propulsion Laboratory |
| LENS | laser-engineered net shaping |
| LEO | low Earth orbit |

| | |
|---|---|
| MJM | multi-jet modeling |
| MSFC | Marshall Space Flight Center |
| MSU | Montana State University |
| | |
| NASA | National Aeronautics and Space Administration |
| NIAC | NASA Innovative Advanced Concepts |
| NRC | National Research Council |
| | |
| R&D | research and development |
| | |
| SBIR | Small Business Independent Research |
| SMD | Science Mission Directorate |
| SSTP | Small Spacecraft Technology Program |
| STMD | Space Technology Mission Directorate |
| | |
| TRL | technology readiness level |
| | |
| UTEP | University of Texas, El Paso |